POLYMER TESTING

NEW INSTRUMENTAL METHODS

POLYMER TESTING

NEW INSTRUMENTAL METHODS

MURALISRINIVASAN SUBRAMANIAN

MOMENTUM PRESS

MOMENTUM PRESS, LLC, NEW YORK

Polymer Testing: New Instrumental Methods
Copyright © Momentum Press®, LLC, 2012

First published by
Momentum Press®, LLC
222 East 46th Street
New York, NY 10017
www.momentumpress.net

ISBN-13: 978-1-60650-242-6 (hard back, case bound)
ISBN-10: 1-60650-242-5 (hard back, case bound)
ISBN-13: 978-1-60650-244-0 (e-book)
ISBN-10: 1-60650-244-1 (e-book)
DOI: 10.5643/9781606502440

Cover design by Jonathan Pennell
Interior design by Derryfield Publishing

10 9 8 7 6 5 4 3 2 1

Printed in the United States of America

Contents

Preface

Global business has caused polymer-related industries to focus strongly on technological competencies. Development of new products is no longer a sequential process leading directly to introduction into the marketplace. Product innovation, development, and in fact the entire process is highly nonlinear and not necessarily sequential. Speeding up and improving the effectiveness of the process must be done in conjunction with a strong regard for safety, health, and environmental values. Energy conservation, product quality, improved process/product economics, and waste reduction are needed while at the same time satisfying and improving the qualitative and quantitative determination of polymeric materials and their properties.

Characterization of polymeric materials often requires the use of instrumental methods. This book provides comprehensive, practical, and up-to-date information about modern instrumental techniques. Only the minimum necessary theory is introduced, as the emphasis is on practical applications. Instead, the book provides in-depth treatment of the use of instrumental methods and includes information that can help determine the choice of instrumental technique.

The most remarkable developments in instrumental methods took place when really high performance became available commercially. These instruments have been in the hands of research scientists for some time now, and they are about to take their place in industry for routine qualitative and quantitative analysis. This book shows the value of these instrumental methods for a wide range of applications within the polymer industries.

Chapter 1 provides an introduction to polymers, their structure, and polymerization techniques. The next four chapters discuss various polymer testing techniques and introduce appropriate instrumental methods: polymer separation techniques in Chapter 2, spectroscopic techniques in Chapter 3, thermal analysis and degradation techniques in Chapter 4, and rheology and other instrumental techniques in Chapter 5.

Chapters 6–8 focus on specific types of polymers, with examples of the use of various instrumental methods to characterize them. Chapter 6 discusses the important and expanding field of thermoplastics. Chapter 7 concentrates on thermosets, their structure, and some major concerns—especially with the concepts necessary to understand the instrumental approach for classification and elucidation. Chapter 8 deals with polymer blends and composites, concentrating on the concepts necessary to understand the properties using instrumental methods.

This book may be of particular interest to researchers in the polymer industry and to those in academia. It provides a valuable overview of recent work in instrumental methods along with some of the significant advances in polymer characterization, together with references to key theoretical papers. Possible trends and future developments in quantitative and qualitative analysis are also discussed.

The objective of this book is to provide useful information about the testing of polymers using new instrumental approaches, and to create a base on which those who are interested may continue their research and studies. Various requirements are covered, with recommendations of instruments for different testing types and methods.

Polymer Testing: New Instrumental Methods discusses both the philosophy and the details of selected instrumentation techniques. We hope that it will encourage scientists and engineers in the polymers field to consider the use of these new approaches to testing, which can be very helpful in evaluating polymer samples with a minimum of time and effort.

Dedication

This book is dedicated to my late father Subramanian, my mother Thangamani, my wife Himachalaganga, and my sons Venkatasubramanian and Sailesh. It is also dedicated to my professors and, above all, to God, who put me on this earth.

Muralisrinivasan Natamai Subramanian

About the Author

Muralisrinivasan Subramanian is a plastics technology consultant specializing in materials, additives, and processing equipment, including troubleshooting. He obtained his B.Sc. in Chemistry from Madurai Kamaraj University and his M.Sc. (1988) in Polymer Technology from Bharathiar University. He received a Post Graduate Diploma in Plastics Processing Technology from CIPET, Chennai. He has also completed his Doctor of Philosophy in Polymer Science from Madurai Kamaraj University. Muralisrinivasan worked in the plastic process industry, mainly in research and development, for 13 years before turning to consultancy and building up an international client base. He teaches plastics processing seminars as well as being a member of the Board of Studies for Expert Colleges in India, dealing with curricula for technology subjects. He is the author of *Update on Troubleshooting in Thermoforming* (Smithers Rapra Technology, 2010), *Basics of Troubleshooting in Plastics Processing: An Introductory Practical Guide* (Wiley-Scrivener, 2011), and *Update on Troubleshooting the PVC Extrusion Process* (iSmithers Rapra Publishing, 2011).

Chapter 1

Introduction

Polymers are complex in nature, and their utility depends on mechanism and process conditions. They become even more complex as a result of blends, composites, and branched and graft structures of unusual architecture. The polymerization must be carefully controlled to obtain the desired properties and processing characteristics. Therefore, it is necessary to understand the influence of polymer properties on their end-use performance. The polymer industry has also grown, and consumption is increasing every year. It is necessary to understand the various facets of the polymerization process in order to understand the variations in polymer properties.

The first useful attempts to create polymers with modification occurred in the middle of the nineteenth century. The nitration of cellulose [1] was reported in 1833. Vulcanization of rubber was patented in 1844 by Goodyear [2]. By the end of the nineteenth century, celluloid was in common use. Celluloid could be said to be the first synthetic or at least partly synthetic plastic. From that point on, the scope of application of many polymers, both natural and synthetic, has been widened by suitable chemical modifications.

Now such polymeric materials are produced in large quantities for commercial applications. Polymers are used in a wide range of applications, such as in the automotive, construction, electronic, cosmetics, and pharmaceutical industries. Polymeric materials can be prepared by various polymerization techniques, including anionic [3], cationic [4,5], or radical [6–9] processes. Polymers have also attracted other-than-conventional materials because of certain advantageous material properties.

However, significant challenges remain in the field of polymers, due largely to major advances in recent years. Particularly in the polymer industry,

DOI: 10.5643/9781606502440/ch1

characterization and quality control remain significant barriers to progress. The challenges in characterization and quality control are to develop experimental techniques for the rapid and precise measurement of properties.

Even in polymer processing, numerous chemical reactions occur, producing considerable changes in physical and chemical properties. Sometimes these reactions may shorten the lifetimes of finished products. During processing with temperature at defined shear rates and in the presence of oxygen, mechanical initiation and thermal oxidation transformations occur. However, at the same time, polymer degradation occurs, and at relevant temperatures chemical reactions may take place due to thermal, mechanical, and autocatalytic factors, with involvement of free radicals, ions, ion pairs, and low-molecular-weight species.

In characterizing polymers, quality control is especially important, with the objective of ensuring that the product remains suitable for its intended end use over its entire lifetime [10].

Traditional methods were manual measurements that were time-consuming, tedious, operator-intensive, with subjective interpretation. To measure certain well-defined properties such as rheology, the common procedure was to use a simple, one-point empirical test method rather than a more complex procedure. Instrumental methods offer operational simplicity, and industrial researchers were quick to accept these more modern techniques. Modern methods have become a preferred analytical tool for general scientific research, and for product specification. Their impact on science and technology has therefore been very significant.

Modern instrumental techniques have provided experimental and theoretical advances in understanding the fundamentals and the properties of polymers in their glassy, rubbery, and molten states. They have helped to classify polymer mechanical properties and have resulted in significant understanding of polymers. Experimental data is also utilized to study theoretical phenomena, which in turn has proved extremely useful to polymer researchers.

Modern instrumental methods are necessary

- To achieve success and be competitive in producing quality products
- To discover the structure and properties of polymeric materials
- For product and process development
- For product identification
- For automation without human error, and at low cost
- To provide data for researchers with minimum effort
- To save time by making the interpretation of results easier

Modern instrumental approaches have been developed to facilitate polymer research and development. They help us understand polymer structures and their relationship to performance. They also establish the validity of techniques for quantitative detection of structural heterogeneity of both polymeric materials and processed end products.

1.1.   Polymer Basics

In a reaction system, monomer or a mixture of monomers is added along with several other ingredients such as a catalyst, an initiator, and water, depending on the type of polymerization process. With agitation at a specific speed, the time, temperature, and pressure are controlled carefully. The monomer is converted into polymer during this process. However, the properties of the final polymer depend on several factors such as the monomer-to-water ratio, if water is present, the degree and speed of agitation, the removal of exothermic heat during the polymerization process, and the solubility of the monomer. For the same type of polymer, different polymerization techniques are also used. A polymer with the same size is difficult to produce, and we are often forced to live with variations in the size and weight of molecules in a polymer. Variations and the weight of the polymer molecules are extremely difficult to control.

1.2.   Morphological Aspects

The polymeric chain as a whole must be considered in terms of its morphological aspects. In a polymeric chain, the architectural arrangements are known as topology. There are many different arrangements, such as linear polymers, networks, branched, cyclic, block, and graft copolymers. However, there are considerable differences in properties between linear polymers versus networks, branched and cyclic polymers, as well as block and graft copolymers. The differences arise primarily because of the different motional constraints imposed and the consequent effects of salvation, entanglement, etc.

1.3.   Chemical Aspects

In polymers, the molecular structure determines chemical aspects. Chemical aspects include types of monomer repeat units and rely ultimately on the solid-state structures and physical properties. The solubility and bulk properties of concern are mainly the glass transition temperature, melting temperature, crystallinity, and modulus. These depend on the chain flexibility, symmetry, and intermolecular attractions, that is, the chemical nature of the polymeric molecules.

1.4.   Classification of Polymers

Polymers have different properties with increasing number of chain ends. Polymer classes, determined by their properties, are discussed below, with the exception of network polymers. A network polymer is one enormous molecule with relatively few chain ends.

1.4.1.   Homopolymers

Homopolymers are prepared using a single type of monomer with a linear chain structure. In the solid state, the behavior of the homopolymer is due to interactions among its molecules. The magnitude of the interactions is dependent on the nature of the intermolecular bonding forces, the molecular weight (number of chain ends), the manner in which the chains are packed, and the flexibility of the polymer.

1.4.2.   Copolymers

When two or more different repeat unit structures are incorporated into a polymeric material, it is known as a copolymer. The linear molecules change the bulk polymer properties by cross-linking. Increasing cross-link density increases melt viscosity and glass transition temperature. Because of their three-dimensional networks, highly cross-linked polymers are insoluble and infusible. One of the most important cross-linking reactions is the vulcanization of rubbers, with reversible elastomeric properties as well as impact strength and shear resistance.

A series of copolymers can be prepared by using varying amounts of each monomer and different techniques, resulting in considerably different properties. The components and the nature of their spatial disposition provide an additional design tool. Copolymers tend to average properties of the constituent monomers in proportion to their relative abundances. Relatively rare copolymers with structural regularity are characterized by combining the properties of the two monomers. Copolymers with alternating structures are capable of interactions which order the structure, which, in turn, can affect their solubility, chemical reactivity, and mechanical properties. The properties and behavior of various arrangements in copolymers, even with the same overall compositions, are distinctly different. This is due to the spatial relationships imposed on the repeat units through covalent bonds, which dictate the conformational properties, interactions with solvents, and intramolecular interactions. Depending on the macromolecular chain, copolymers can be classified with respect to the sequence of monomeric units.

1.4.3.   Block Copolymers

Through chemical combination of two different monomers, A and B, either block copolymers of structurally distinct polymers or graft copolymers with suitable linking chain ends can be produced. Block copolymers contain relatively long separate sequences of two or more different repeat units [11,12]. Chemically different sequences or blocks are usually not mutually miscible and hence separate by phases. Such microdomain formation results in a material that has some of the properties of the separate blocks.

Figure 1.1 shows block diagrams of (1) a block copolymer and (2) a graft copolymer. The importance of block copolymers can be realized from the fact that, depending on the arrangement of the blocks, their compatibility, and their melting and glass transition temperatures, random copolymers can be produced using high-tensile-strength materials in a relatively larger temperature window. Such block copolymers show microphase separation, in which glassy or crystalline blocks tend to segregate into domains which

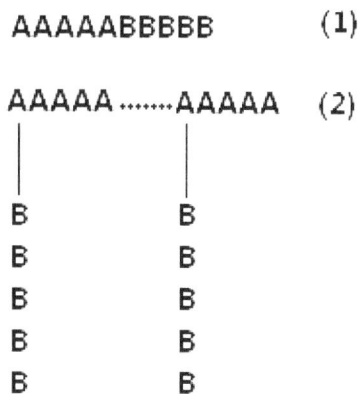

```
         AAAAABBBBB         (1)

         AAAAA ........ AAAAA    (2)
         |            |
         |            |
         B            B
         B            B
         B            B
         B            B
         B            B
```

Figure 1.1.   (1) Block copolymer. (2) Graft copolymer.

serve to anchor elastomeric blocks and act as effective physical cross-linking points. These physical cross-linking points are reversible. Examples of such thermoplastic elastomers include polyurethanes and polystyrene-polydiene block copolymers [13].

1.4.4. Graft Copolymers

Graft copolymers are branched polymers that are structurally different from the primary or backbone chain (graft base). Comb-shaped graft polymers contain branches of generally equal length attached more or less equidistantly along the main chain. Various graft copolymers can be obtained, depending on compatibility and crystallization tendencies, with possible improvement in properties that depend on phase-separation morphology, analogous to the block copolymers of the graft and the base. One of the important applications of graft copolymers is grafting with polar units to produce hydrophilic materials.

1.4.5. Cyclic Polymers

In the broad spectrum of polymer classification, cyclic polymers represent an extreme case, since they have no chain ends. Cyclic macromolecules have aroused considerable interest because of their special dilute solution, diffusional, and rheological properties [14,15]. Cyclic polymers have properties which include lower hydrodynamic volumes, melt viscosities, and coefficients of friction, with higher glass transition temperatures. To prepare well-defined cyclic polymers, anionic polymerization has been utilized [16].

1.4.6. Branched Polymers

In branched polymers, the main chain is typically the longest element, with either short or long pendant chains. In series branching the main chains are themselves branched. Dendritic structure has very extensive and precisely defined series branching [17].

Polyethylene is a good example of a branched polymer [18]. The density and melting point are higher than for linear polyethylene. If the chain is substantially branched, the packing efficiency is reduced and the crystalline content is lowered. This provides a good handle on lowering the crystalline content of the polymer. The glass transition temperature (T_g) may be significantly affected, depending on the extent of branching. Even a small number of branches on a polymer chain will reduce the T_g of the polymer due to increased free volume. However, with more branching, the density has the same effect as side groups in restricting mobility and hence means a higher T_g.

1.4.7. Dendrite

The regular dendrite structure, which possesses branching with radial symmetry, has received substantial interest. Three distinguishing structural features are available from dendrites: a core, interior layers with repeating units that are radially attached to the core, and an exterior or surface of terminal functionality attached to the outermost generation. The properties and applications of dendrimers are still being investigated. Properties such as controllable size, compartmentalization, preorientation, cage localization, solubilization, polarity, and surface group/counter-ion effects should be of considerable interest in the quest for novel catalytic and delivery media [19–29].

1.4.8. Star-Shaped Polymers

Star-shaped polymers have branches of almost equal length radiating out from a single branch point and are characterized by low hydrodynamic volume due to their low radius of twist. In the solution properties of the polymer, the lower intrinsic viscosity depends on the number of branches relative to linear polymers of the same molecular weights. The branches in the polymer can be homopolymeric or diblock copolymeric [19–29]. These polymers can be prepared by anionic polymerization [30,31]. Figure 1.2 shows the polymer topologies of polymer structures [32].

1.4.9. Cross-links

In polymeric systems, the presence of chemical cross-links has the effect of increasing T_g when the density of cross-links is higher in the range of the transition region. The range broadens, and the T_g may not be observed at all. Cross-linking tends to reduce the specific volume of the polymer, because the free volume is reduced and molecular motions become more difficult, so T_g is raised. The polymer is cross-linked with improvement in dimensional stability, resistance to solvents, and heat stability. The cross-linked

I. Cyclic Polymers (No Chain Ends)

II. Linear Polymers (Two Chain Ends)
   Homopolymer
   Random Copolymer
   Alternating Copolymer
   Block Copolymer

III. Branched Polymers
   (More than Two Ends)

   Star Polymers (Homo- or Block)

   Randomly Branched Polymers

   Graft/Comb Polymers
   (Long Chain- &
   Short Chain - Branching

   Dendritic

IV. Network Polymers
   (One Enormous Molecule with few chain ends)

   Interpentrating
   Polymer
   Network

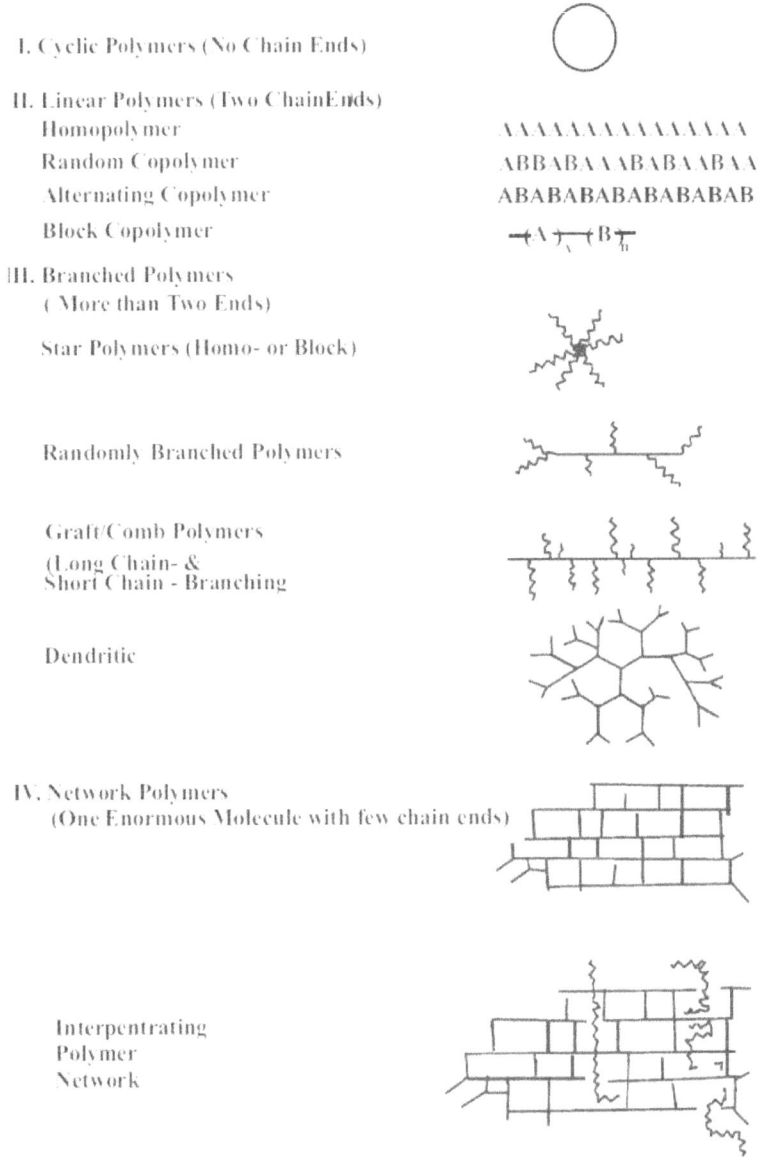

Figure 1.2. Polymer topologies. [Reprinted from H. W. Gibson, M. C. Bheda, and P. T. Engen, *Prog. Polym. Sci.* 19, 843 (1994) with permission from Elsevier Science Ltd. Copyright © 1994 Elsevier Ltd. All rights reserved.]

thermosetting resins are characterized by high modulus, low damping, and low creep rate [33]. Table 1.1 shows some pairs of functional groups and the resulting cross-linking [34].

Table 1-1 Some Pairs of Functional Groups and the Resulting Cross-linking

Functional group A	Functional group B	Cross-linking formed
—COOH	—C—C (epoxide, O)	—COO—C—C— OH
—COOH	—C (N—C, O—C)	—COO-C-C-N-C (=O)
—COOH	—N (C, C)	—COO—C—C—NH—
—NH	—C—C (epoxide, O)	N—C— (=O)
NH	—COOH	N—C— (=O)
NH	—C (N—C, O—C)	N—C—C—NH—C— (=O)

Source: Reprinted from Q. Wang, S. Fu, and T. Yu, *Prog. Polym. Sci.* 19, 703 (1994) with permission from Elsevier Science Ltd. Copyright © 1994 Elsevier Ltd. All rights reserved.

1.4.10.   Interpenetrating Polymer Networks

Interpenetrating polymer networks (IPNs) are formed by polymerization of one or more monomers or pairs of monomer in the presence of a preformed cross-linked polymer or by simultaneous but chemically independent formation of two or more cross-linked polymers [35]. Because of topological constraints imposed by the cross-links, IPNs are insoluble and exhibit little creep. Starting from at least one monomer and using rapid polymerization reactions, phase separation is not as extensive as in blends of preformed homopolymers [36].

1.5.   Polymerization Techniques

1.5.1.   Free-Radical Polymerization

The free-radical process was developed originally during World War II as part of the U.S. effort in support of its large synthetic rubber industry.

More recently, today's rapidly developing science of anionic or coordinated polymerization has had little impact on free-radical systems. Improved polymerization techniques have been developed, but the basic information on mechanisms has not been greatly added to. A brief account of the major features of free-radical–initiated polymerizations is appropriate to putting the whole field of polymerization into perspective.

In free-radical solution polymerization, a solvent and an initiator are used. The jacketed reactor is provided with a stirrer for thorough mixing of the reactants. By circulating water at an appropriate temperature through the reactor jacket, heating and cooling of the reaction mixture is achieved. The reactor temperature is controlled by a cascade control system consisting of proportional integral derivative (PID) controllers. The manipulated variables for the controllers are hot-water and cold-water flow rates. The hot and cold water streams are mixed before entering the reactor jacket and provide heating and cooling for the reactor. The jacket outlet temperature is fed back to the controllers. A detailed mathematical model covering reaction kinetics and heat and mass balances was developed and validated using real reactor operation data [37,38]. Figure 1.3 shows conventional free-radical reactions [39].

The molecular-weight distribution (MWD) generated in a free-radical polymerization is determined by the relative rates of initiation, propagation, termination, and chain transfer. In most cases, the chain length of the polymer is fixed for the majority of chains once the propagating chain end has been involved in a termination or transfer event.

Chain transfer to polymer normally affects a small fraction of polymer chains in the system. On the other hand, it is well known that in stepwise polymerization, a significant amount of reshuffling of the chain length distribution occurs under appropriate reaction conditions due to interchange reactions [40–42]. This reshuffling of the chain length distribution of linear polymers ideally results in the MWD approaching the most probable distribution [36,40,43].

1.5.2. Addition Polymerization

Addition polymerization is usually accompanied by side reactions such as chain transfer and termination reactions. It leads to an irreversible deactivation of the growing species. Consequently, polymers with multimodal or broad MWD are formed. Using selected initiators under carefully optimized conditions, the possibility of side reactions can be diminished so that the polymer molecules retain their ability to grow even after the monomer is completely polymerized. Therefore, when further monomer is added to the

1. Initiation

$$\tfrac{1}{2}\,In_2 \xrightarrow{\ K_{diss}\ } In_{\bullet}$$

$$In_{\bullet} + M \xrightarrow{\ K_i\ } P_{1\bullet}$$

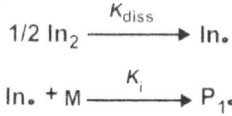

2. Propagation

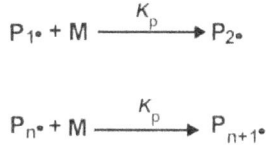

$$P_{1\bullet} + M \xrightarrow{\ K_p\ } P_{2\bullet}$$

$$P_{n\bullet} + M \xrightarrow{\ K_p\ } P_{n+1\bullet}$$

3. Chain transfer

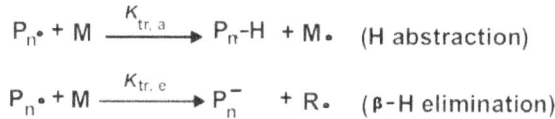

$$P_{n\bullet} + M \xrightarrow{\ K_{tr,\,a}\ } P_n\text{-H} + M_{\bullet} \quad \text{(H abstraction)}$$

$$P_{n\bullet} + M \xrightarrow{\ K_{tr,\,e}\ } P_n^- + R_{\bullet} \quad (\beta\text{-H elimination})$$

4. Chain termination

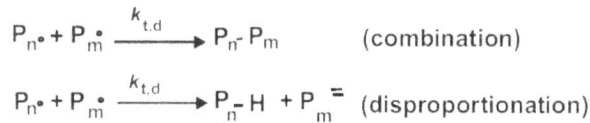

$$P_{n\bullet} + P_{m\bullet} \xrightarrow{\ k_{t.d}\ } P_n\text{-}P_m \qquad \text{(combination)}$$

$$P_{n\bullet} + P_{m\bullet} \xrightarrow{\ k_{t.d}\ } P_n\text{-H} + P_m^{=} \quad \text{(disproportionation)}$$

Figure 1.3. Elementary reactions in conventional free-radical polymerization. [Reprinted from F. di Lena and K. Matyjaszewski, *Prog. Polym. Sci.* 35, 959, 1021 (2010) with permission from Elsevier Science Ltd. Copyright © 2010 Elsevier Ltd. All rights reserved.]

polymerization system, second-stage polymerization ensues, resulting in an increase in polymer molecular weight.

1.5.3. Anionic Polymerization

Recent developments in anionic polymerization [44,45] offer significant opportunities for enhancing the utility of styrene polymers. The ability to form random and block copolymers and high reaction rates even for high-molecular-weight resins opens up possibilities for the future, such as copolymers of styrene and α-methyl styrene for high-heat applications.

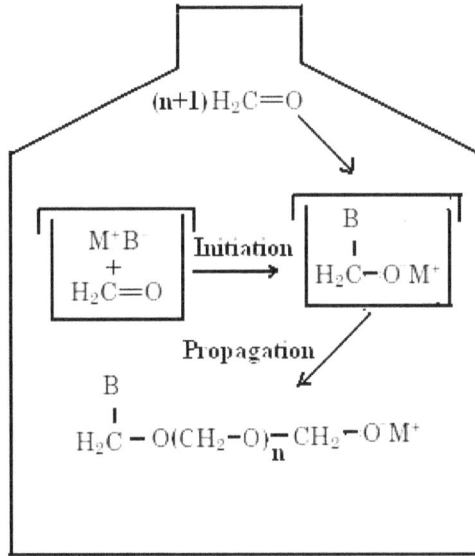

Figure 1.4. Initiation and propagation in anionic polymerization.

Figure 1.4 illustrates the initiation and propagation reactions in anionic polymerization. Initiators are usually weak bases such pyridines. The exact structure of M+ and B− depends on the solvent. Further addition by chain transfer helps to regulate molecular weight. Termination is commonly similar to that of vinyl polymers.

1.5.4. Cationic Polymerization

Cationic polymerization is characterized by growth by the addition of monomer to the cationic chain end, as shown in the following equations [47]:

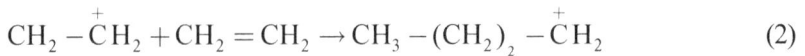

$$H^+ + CH_2 = CH_2 \rightarrow CH_2 - \overset{+}{C}H_2 \tag{1}$$

$$CH_2 - \overset{+}{C}H_2 + CH_2 = CH_2 \rightarrow CH_3 - (CH_2)_2 - \overset{+}{C}H_2 \tag{2}$$

$$CH_3 - (CH_2)_2 - \overset{+}{C}H_2 + CH_2 = CH_2 \rightarrow CH_3 - (CH_2)_4 - \overset{+}{C}H_2 \tag{3}$$

Traditionally, cationic polymerization has been considered as polymerization in which either a free solvated cation or the electron-deficient portion of an ion pair is the point of propagation. It is based on the inhibition of the polymerization by nucleophiles. Color formation has been observed sometimes due to free ions present in the process, especially in styrene polymerizations, which makes propagation by cations likely. Conductance measurements, however, do not necessarily show that cations are the carriers of the polymerization.

Modern cationic polymerization began about 20 years ago with the discovery of cocatalysis. In the 1950s, research was being conducted by relatively few scientists. The early 1960s, however, brought new life to investigations of cationic polymerization. Advances in several related fields [36,46] have also influenced the progress in cationic polymerization. The major contributions include detailed knowledge of anionic polymerization, living polymer systems, ion pairs, solvation of ions, carbonium ions, and solid-state polymerization. The monomers form the first monomer cation by taking up a cation from the iniator system.

1.5.5.   Ring-Opening Polymerization

Ring-opening polymerization (ROP) is a useful method for preparing microstructure-controlled polymers and polymers with novel structures, which may not be possible using conventional polymerization techniques. The initiation of ROP can be achieved by thermal, anionic, or transition-metal catalysts. Ionically catalyzed ROP has been particularly well investigated. Currently, thermally induced ROP is the most generally used polymerization method [48].

Ring-opening polymerization is capable of introducing polar segments into polyolefin main chains. The most common segments incorporated into polyolefins are poly(ϵ-caprolactone), poly(L,L-lactides), poly(propylene glycol), and so on. To obtain these block or graft copolymers through ROP, polyethylene-bearing hydroxyl groups are generally a necessary precursor. Free-radical ring-opening polymerizations have received considerable attention as a means of reducing the level of shrinkage normally associated with free-radical polymerization [49–54].

1.5.6.   Chain-Transfer Polymerization

In-situ chain transfer during the polymerization is an effective method for introducing functional groups to polymer chains and yield terminally functionalized polyolefins [55–57]. Several compounds, including organoborane [58,59], alkyl aluminum [60,61], and alky zinc [62–64], are reported to be effective chain-transfer agents in olefin polymerizations.

1.6.   Polymerization Processes

1.6.1.   Bulk Polymerization

Monomers do not readily polymerize in bulk or solution with free-radical catalysts. The rates are slow and the polymers are either of low molecular weight or cross-linked and insoluble. Bulk polymerization has not been long studied in detail, and as the monomers available were not always of the highest purity, earlier results may be unreliable. Thermal polymerizations are slow, and even at temperatures of 100°C, times in excess of 100 h are required for good yields of polymer. Addition of benzoyl peroxide accelerates the polymerizations to some extent, but this compound is rapidly decomposed and a more stable type of initiator such as diazoaminobenzene is more effective.

Polymers should be totally insoluble in common solvents to avoid losses and to facilitate easy handling, separation, and purification. They should have a high degree of substitution with uniformity. To reduce undesirable competing branching, monomers should undergo straightforward reactions.

Polymers should be capable of being recycled. The degree of cross-linking normally determines the polymer effective pore size. A high porosity allows good flow properties in the polymer. The backbone of the polymer should be inert toward most chemicals, and the structure of the backbone should be chemically stable and not susceptible to degradative scission by chemical reagents under ordinary conditions.

Linear polymers which are soluble in monomer as reagents in polymeric reactions may have a disadvantage in the form of difficulty of separation from the reaction mixture.

1.6.2.   Emulsion Polymerization

Emulsion polymerization involves a propagation reaction of free radicals with monomer molecules in a very large number of discrete polymer particles dispersed in a continuous aqueous phase. This process is widely used to produce waterborne resins with various colloidal and physicochemical properties. This heterogeneous free-radical polymerization process involves emulsification of a relatively hydrophobic monomer in water by an oil-in-water emulsifier, followed by an initiation reaction with either a water-insoluble initiator or an oil-soluble initiator [65–70].

Emulsion polymerization was developed between 1930 and 1950, and it has become a very successful industrial process. The process is carried out in a heterogeneous system, commonly with an aqueous and a nonaqueous phase. It is a compartmentalized polymerization reaction taking place in a large number of reaction monomers dispersed in a continuous external phase. The polymer and hence the monomer are usually part of the nonaqueous phase.

A typical emulsion polymerization system consists of water, monomer(s), emulsifier (surfactant), initiator, modifier, and other additives. Usually, the monomer is sparingly soluble in water and generates a water-insoluble polymer which is swollen by monomer. Emulsification initially results in micelles swollen with solubilized monomer and surfactant-stabilized monomer droplets. As initiator decomposition takes place, a new phase appears, latex particles, which contain macromolecules of a fairly high degree of polymerization, and are swollen with monomer and stabilized by surfactant [34].

Emulsion polymerization involves a complex heterogeneous chemical reaction. The formulation and process parameters strongly impact the characteristics of the resulting polymer particles. Formulation variables include the types and amounts of the monomers used in the polymerization (principal monomers, functional monomers, cross-linkers), the type and concentration of the initiator (thermal, or redox water-soluble or oil-soluble), and the type and concentration of surfactant or stabilizer (anionic, nonionic, cationic, zwitterionic, polymeric, or protective colloid). In addition, small quantities of rheological modifiers, chain-transfer agents, buffers, biocides and fungicides, antioxidants, and UV absorbers may be added [71].

Process variables may include the type and configuration of the reactor, the polymerization process (batch, semicontinuous, continuous, shotgrowth, seeded), the feed strategy [of monomer (s), initiator, surfactant], the type of impeller (e.g., fluid foil, Rushton turbine, paddle), agitation rate, and process temperature. The situation is even more complex because many of these variables interact with one another to influence the final product.

Mechanisms of particle nucleation in emulsion polymerization may include micellar, homogeneous, droplet, or combinations of these mechanisms. The type of mechanism is dependent on the water solubilities of the monomers used in the polymerization; the type of initiator; the type, concentration, and partitioning of the surfactant used in the formulation; and the state of the monomer dispersion (droplet size and stability) at the polymerization temperature. The rate of polymerization is critical; these polymerizations are exothermic, and a reactor must be able to safely remove the heat of reaction. The rate is controlled by the polymerization temperature, initiator and surfactant concentrations (influencing the number of particles that are generated), and other factors. The polymer particle size and size distribution depend on these same variables; narrow distributions are achieved when all the particles are all nucleated in a relatively short time and then grow uniformly [72–74].

1.6.3.   Solution Polymerization

Solution polymerizations have been carried out in traditional organic solvents or in aqueous media with monomer. Emission of volatile organic compounds and polluted aqueous waste streams into the environment has prompted a search for less harmful solvent alternatives. Carbon dioxide has intrinsic environmental advantages of being nontoxic, nonflammable, and easily separated and recycled. It acts as a good solvent under relatively high pressures in the liquid or supercritical state. It is also widely available and inexpensive, which suggests a potential for large-scale industrial use [75].

1.6.4.   Suspension Polymerization

In suspension polymerization, the polymer precipitates in the monomer droplet. Suspension polymerization is a series of processes involving emulsifying monomers to droplets with stirring in a suspending medium. Usually, the free-radical initiator is soluble in the monomer, which is insoluble in suspension polymerization. The polymer formed is soluble in the monomer. The polymer forms as nonporous, spherical beads. Therefore this process is termed as suspension polymerization. Polymer precipitates are also composed of many smaller primary particles. They are opaque, usually possess an irregular surface, and may have internal porosity. Figure 1.5 shows a schematic of suspension polymerization.

1.6.5.   Step-Growth Polymerization

Step-growth polymerization involves reactions of functional groups of monomers in a stepwise progression from dimers, trimers, etc. This progression will eventually form high polymer. Figure 1.6 shows a simple laboratory technique for step-growth polymerization.

The process is accompanied by elimination of small molecules during the reaction step. Hence it is called polycondensation. Unlike other reactions of polymers such as polyolefins and vinyl polymers, this process can lead to different types of monomer attachments with associated mechanical and other property changes. The incorporation of new monomers into the backbone is required to produce the property changes.

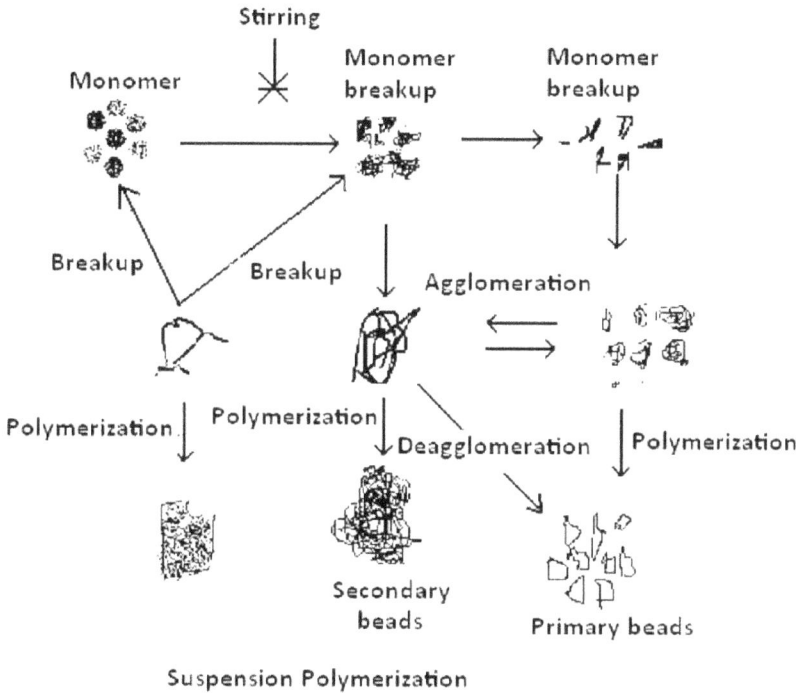

Figure 1.5.   Schematic of suspension polymerization.

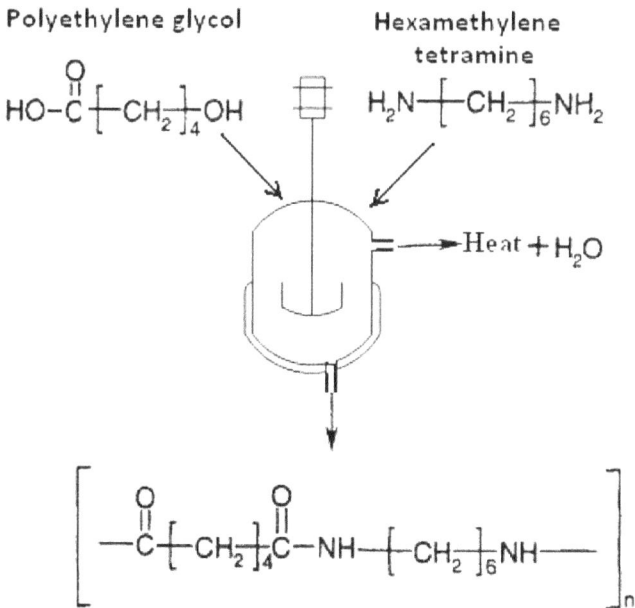

Figure 1.6.   Simple step-growth polymerization laboratory technique.

Step-growth polymers find applications in many engineering resins, but development and commercialization of new step-growth monomers and polymers are still challenging tasks.

1.7.   Polymer Synthesis

Polymer synthesis or modification suitable for different applications lies in the versatility of the synthesis from chemicals and the wide range of reactions possible. The chemicals are connected to different molecular entities via monomers. Polymer synthesis affords a product with particular properties which can be utilized in a variety of ways. Therefore, polymers belong to a very wide series of same and/or different monomers. Polymers can confer on the monomer molecule typical basic properties such as the possibility of chains that are linear or branched or both.

However, polymers are quite different from the monomer from which they are derived, largely because of

- Steric hindrance by the polymer to the attachment of the monomeric chain
- Adsorption of monomer onto the polymeric material
- The need for drastic conditions to reach satisfactory conversion of monomer into polymer

To achieve maximum advantages of polymers, it is important to minimize any potential problems.

Much progress has made with respect to polymer synthesis, characterization, and structure–property correlations. This progress has yielded better understanding and has brought a wide range of potential applications into consideration with respect to instrumental analysis. Both synthesis and characterization of polymeric materials by instrumental methods has developed to explore structure–property relationships. These methods have also led to new syntheses, concepts, and architectures, and ultimately many new applications.

Besides the behavior of the polymer, the functional groups present in the substrate and monomer also show peculiar properties which derive from their own structure. Of these, the possibility of monomer substitution by a functional group is an important aspect. It is noteworthy that several reactions of macromolecular chemistry involve monomers and their chemistry.

For example, hardening of novolacs by hexamethylenetetramine [76] involves both aminomethylation and substitution reactions as well as a catalytic effect. Synthesis and chemical properties of monomers are also present in the chemistry of aniline [77] and in the treatment of protein molecules with formaldehyde, such as hardening of casein [77]. The synthesis of urea resin and related products uses an amidomethylation similar to novalocs.

In polymer synthesis, the major requirements are

1. Ease of preparation with a controlled degree of polymerization
2. Compatibility with most organic reagents
3. Stability with respect to chemical and mechanical properties
4. Inertness of the backbone of the polymer toward reactants and regents
5. Low cost and commercial availability of the monomer

1.8.   Polymer Structure and Properties

In polymers, a monomer or repeating chemical unit is the common feature. The simplest polymer is usually a linear construction of monomers. However, polymers may also be made up of a small number of different chemical units. They may have branches or even a network instead of a simple linear chain. Yet a polymer molecule is a rather linear system with strong covalent bonds in the chain direction and only weak coupling laterally. The constitution and degree of polymerization strongly influence the conformation of polymers.

In polymers, the actual conformation determines the packing of material in the condensed state, which will be either amorphous, melt or glass, or partially crystalline. The possibilities include the set of all arrangements that can be formed by rotation around the chemical bonds. Increasing polydispersity increases the melt strength of a resin, mainly due to increase of its elasticity [78–82]. However, this is only partially observed in industrial practice.

Different polymerization techniques used in different industries result in polymers with different molecular architectures, making their rheological and processing characteristics technology-dependent. The number and size of the resulting polymer molecules may be affected significantly by the rate of polymerization, the solvent, and the other media involved in each technique. These factors cause significant differences in the molecular architecture and characteristics, which thus obviously affects the properties of the polymers.

The glass transition temperature is a very important physical parameter of polymers and is directly related to their physical character. The geometric and structural shapes of a macro are dependent on its composition. Various ingredients and groups present in a polymer influence the physical properties of the polymer material. Glass transition temperature is thus reflective of the geometric and structural shapes of the polymer [83,84].

Polymer properties (e.g., mechanical, thermal, and structural) can be analyzed by a variety of characterization tools, which allows the determination of structure–property relationships [85,86]. Today, advanced characterization tools such as gel permeation chromatography (GPC), nuclear magnetic resonance (NMR) spectroscopy, nanoindentation, thermogravimetric analysis, or contact-angle measurements are routinely used for the determination of several important polymer properties. To facilitate progress in this field, both the synthesis of new classes of polymers and the development of new synthetic methods for polymers are indispensible [87].

1.9.   Requirements for Instrumental Methods

Polymer science has achieved major advances, but significant challenges remain. Continued development of experimental techniques is important for further progress. Development of polymer industries and new product development require thorough characterization of potential materials. However, polymer characterization and quality control remain significant barriers to rapid and precise measurements of properties. Therefore, polymer characterization, analysis, and testing is required at every stage from polymer synthesis to product performance.

Polymer testing can be expected to take on an even greater role in synthesis and development in the coming years. Testing is required to improve the effectiveness and to characterize complex polymeric materials. It is necessary to combine analysis and characterization of polymers to establish structure, property, and morphology relationships that can form the basis of knowledge. Polymer testing is necessary to guard against variation in properties and processing, which may otherwise be very costly.

References

1. H. Braconnet, *Ann. Chim. Phys.* 52, 290 (1833).
2. U.S. Patent 3,633.
3. N. Hadjichristidis, M. Pitsikalis, S. Pispas, and H. Iatrou, *Chem. Rev.* 101, 3747 (2001).
4. O. W. Webster, *Science* 496, 887 (1991).
5. E. J. Goethals, M. Dubreuil, Y. Wang, I. De Witte, D. Christova, S. Verbrugghe, N. Yanul, L. Tanghe, G. Mynarczuk, and F. Du Prez, *Macromol. Symp.* 153, 209 (2000).

6. K. A. Davis and K. Matyjaszewski, *Adv. Polym. Sci.* 159, 2 (2002).
7. K. Matyjaszewski and J. Xia, *Chem. Rev.* 101, 2921 (2001).
8. C. J. Hawker, A. W. Bosman, and E. Harth, *Chem. Rev.* 101, 3661 (2001).
9. S. Perrier and P. Takolpuckdee, *J. Polym. Sci. A* 43, 5347 (2005).
10. J. M. Dealy and P. C. Saucier, *Rheology as a Tool for Quality Control in the Plastics Industry,* Hanser-Gardner Publications, Cincinnati, OH (1999).
11. G. Reiss, G. Hurtrez, and P. Bahadur, in *Encyclopedia of Polymer Science and Engineering,* 2nd ed., Vol. 2, pp. 324–434, John Wiley, New York (1985).
12. D. J. Meier (ed.), *Block Copolymers: Science and Technology,* MMI Symposium Series, MMI Press/Harwood; Academic Press (1983).
13. J. M. G. Cowie, *Polymers—Chemistry and Physics of Modern Materials,* p. 271, Intertext Books, London (1973).
14. J. A. Semylen (ed.), *Cyclic Polymers,* Elsevier, New York (1986).
15. G. B. Mckenna, B. J. Hostetter, N. Hadjichristidis, L. J. Fetters, and D. J. Plazek, *Macromolecules* 22, 1834 (1989).
16. J. Roovers and P. M. Toporowski, *Polym. Sci., Polym. Phys.* 26, 125 (1988).
17. H. G. Elias, *Macromolecules,* chap. 34, p. 50, Plenum Press, New York (1984).
18. J. M. G. Cowie, *Polymers—Chemistry and Physics of Modern Materials,* p. 193, Intertext, London (1973).
19. P. G. De Gennes and H. J. Hervert, *Phys. Lett.* 44, 351 (1983).
20. M. J. Maciejewski, *Macromol. Sci., Chem. A* 17, 689 (1982).
21. D. A. Tomalia, D. M. Hedstrand, and L. R. Wilson, in *Encyclopedia of Polymer Science and Engineering,* 3rd ed., pp. 46–92, John Wiley, New York (1990).
22. D. A. Tomalia, A. M. Naylor, and W. A. Goddard III, *Angew. Chem. Int. Ed. Engl.* 29, 138 (1990).
23. G. R. Newkome, C. N. Moorefield, G. R. Baker, R. K. Behera, G. H. Escamillia, and M. J. Saunders, *Angew. Chem, Int. Ed. Engl.* 31, 917 (1992).
24. H. B. Mekelburger, W. Jaworek, and F. Vogtle, *Angew. Chem. Int. Ed. Engl.* 31, 1571 (1992).
25. G. R. Newkome, C. N. Moorefield, and G. R. Baker, *Aldrichim. Acta* 25(2), 31 (1992).
26. D. A. Tomalia and H. D. Durst, *Top. Curr. Chem.* 165, 193 (1993).
27. Z. Xu and J. S. Moore, *Angew. Chem. Int. Ed. Engl.* 32, 1354 (1993).
28. E. M. M. De Brabander-Vandenberg and E. W. Meijer, *Angew. Chem. Int. Ed. Engl.* 32, 1308 (1993).
29. I. Grrsov and J. M. J. Frechet, *Macromolecules* 26, 6536 (1993).
30. J. E. McGrath (ed.), *Anionic Polymerizations,* ACS Symposium Series No. 166, American Chemical Society, Washington, DC (1981).
31. M. Van Beylen, S. Bywater, G. Smets, M. Szwarc, and D. J. Worsfold, *Adv. Polym. Sci.* 86, 87 (1988).
32. H. W. Gibbson, M. C. Bheda, and P. T. Engen, *Prog. Polym. Sci.* 19, 843 (1994).
33. J. M. G. Cowie, *Polymers—Chemistry and Physics of Modern Materials,* p. 283, Intertext, London (1973).
34. Q. Wang, S. Fu, and T. Yu, *Prog. Polym. Sci.* 19, 703 (1994).

35. L. H. Sperling, *Interpenetrating Polymer Networks and Related Materials,* Plenum Press, New York (1981).
36. P. J. Flory, *Principles of Polymer Chemistry,* Cornell University Press, Ithaca, NY (1953).
37. D. Achilias and C. Kiparissides, *Macromolecules* 25, 3739 (1992).
38. A. Penlidis, S. R. Ponnuswamy, C. Kiparissides, and K. F. O'Driscoll, *Chem. Eng. J.* 50, 95 (1992).
39. F. di Lena and K. Matyjaszewski, *Prog. Polym. Sci.* 35, 959, 1021 (2010).
40. P. J. Flory, *J. Am. Chem. Soc.* 64, 2205 (1942).
41. S. Collins, S. K. Peace, R. W. Richards, W. A. MacDonald, P. Mills, and S. M. King, *Polymer* 42, 7695 (2001).
42. P. B. Zetterlund, R. G. Gosden, and A. F. Johnson, *Polym. Int.* 52, 104 (2003).
43. G. Odian, *Principles of Polymerization,* 4th ed., sec. 2–7c, John Wiley, Hoboken, NJ, (2004).
44. D. B. Priddy and M. Pirc, *J. Polym. React. Eng.* 1 343 (1993).
45. R. Thiele, *Chem. Eng. Technol.* 17, 127 (1994).
46. G. A. Olah and P. R. Schleyer, *Carbonium Ions,* Interscience, New York (1968).
47. G. Heublein, Zum Ablauf ionischer Polymerisationsreaktionen, p. 73, Akademieverlag, Berlin (1975).
48. T. Baumgartner, F. Jakle, R. Rulkens, G. Zech, A. J. Lough, and I. Manners, *J. Am. Chem. Soc.* 124, 10062 (2002).
49. E. Klemm and T. Schulze, *Acta Polym.* 50, 1 (1999).
50. F. Sanda and T. Endo, *J. Polym. Sci. A* 39, 265 (2001).
51. R. A. Evans, G. Moad, E. Rizzardo, and S. H. Thang, *Macromolecules* 27, 7935 (1994).
52. R. A. Evans and E. Rizzardo, *Macromolecules* 29, 6983 (1996).
53. R. A. Evans and E. Rizzardo, *J. Polym. Sci, A* 39, 202 (2001).
54. T. Endo and T. Yokozawa, in W. J. Mijs (ed.), *New Methods for Polymer Synthesis,* Plenum Press, New York (1992).
55. P. F. Fu and T. J. Marks, *J. Am. Chem. Soc.* 117, 10747 (1995).
56. G. Xu and T. C. Chung, *J. Am. Chem. Soc.* 121, 6763 (1999).
57. S. B. Amin and T. J. Marks, *Angew. Chem. Int. Ed.* 47, 2006 (2008).
58. T. C. Chung, G. Xu, Y. Y. Lu, and Y. L. Hu, *Macromolecules* 34, 8040 (2001).
59. W. T. Lin, J. Y. Dong, and T. C. Chung, *Macromolecules* 41, 8452 (2008).
60. Z. G. Cai, M. Shigemasa, Y. Nakayama, and T. Shiono, *Macromolecules* 39, 6321 (2006).
61. F. Rouholahnejad, D. Mathis, and P. Chen, *Organometallics* 29, 294 (2010).
62. G. J. P. Britovsek, S. A. Cohen, V. C. Gibson, P. J. Maddox, and M. van Meurs, *Angew. Chem. Int. Ed.* 41, 489 (2002).
63. D. J. Arriola, E. M. Carnahan, P. D. Hustad, R. L. Kuhlman, and T. T. Wenzel, *Science* 312, 714 (2006).
64. A. Xiao, L. Wang, Q. Q. Liu, H. Yu, J. Wang, J. Huo, Q. Tan, J. Ding, W. Ding, and A. M. Amin, *Macromolecules* 42, 1834 (2009).
65. F. A. Bovey, I. M. Kolthoff, A. I. Medalia, and E. J. Meehan, *Emulsion Polymerization,* Interscience, New York (1965).

66. D. C. Blakely, *Emulsion Polymerization. Theory and Practice,* Applied Science, London (1975).
67. V. I. Eliseeva, S. S. Ivanchev, S. I. Kuchanov, and A. V. Lebedev, *Emulsion Polymerization and Its Applications in Industry,* Consultants Bureau, New York (1981).
68. J. Barton and I. Capek, *Radical Polymerization in Disperse Systems.* Ellis Horwood, New York (1994).
69. R. G. Gilbert, *Emulsion Polymerization: A Mechanistic Approach,* Academic Press, London (1995).
70. R. M. Fitch, *Polymer Colloids: A Comprehensive Introduction,* Academic Press, London (1997).
71. A. Klein, and E. S. Daniels, in P. A. Lovell and M. S. El-Aasser (eds.), *Emulsion Polymerization and Emulsion Polymers,* pp. 207–237, John Wiley, Chichester, U.K. (1997).
72. J. M. Asua and H. A. S. Schoonbrood, *Acta Polym.* 49, 671 (1998).
73. A. Guyot, *Polymerizable Surfactants,* Surfactant Science Series 74 (Novel Surfactants), pp. 301–332, Marcel Dekker, New York (1998).
74. A.-C. Hellgren, P. Weissenborn, and K. Holmberg, *Prog. Org. Coat.* 35, 79 (1999).
75. M. A. McHugh and V. J. Krukonis, *Supercritical Fluid Extraction,* 2nd ed., Butterworth-Heinemann, Boston (1994).
76. K. J. Saunders, *Organic Polymer Chemistry,* pp. 286 and 295, Chapman & Hall, London (1977).
77. K. J. Saunders, *Organic Polymer Chemistry,* p. 316, Chapman & Hall, London (1977).
78. K. J. Saunders, *Organic Polymer Chemistry,* p. 375, Chapman & Hall, London (1977).
79. J. M. Dealy and K. F. Wissbrun, *Melt Rheology and Its Role in Plastics Processing,* Reinhold, New York (1995).
80. D. V. Rosato and D. V. Rosato (eds.), *Blow Molding Handbook,* Hanser, New York (1989).
81. J. M. Charrier, *Polymeric Materials and Processing,* Hanser, New York (1990).
82. F. P. La Mantia and D. Acierno, *Polym. Eng. Sci.* 25, 279 (1985).
83. K. M. Idriss Ali and T. Sasaki, *Radiat. Phys. Chem.* 43, 371 (1994).
84. M. Azam Ali, M. A. Khan, and K. M. Idriss Ali, *Polym. Plast. Technol. Eng.* 34, 447 (1995).
85. D. Tyagid, J. L. Hedrick, D. C. Webster, J. E. Mcgrath, and G. L. Wilkes, *Polymer* 29, 883 (1988).
86. R. Hoogenboom, *Macromol. Chem. Phys,* 208, 18 (2007).
87. Y. Imai, *React. Funct. Polym.* 30, 3 (1996).

Chapter 2

Polymer Separation Techniques

Polymers have molar mass, chemical composition, functionality, and molecular architecture, and there is high demand on their distributed properties. The performance of polymer materials is due to their properties. It is therefore necessary to have detailed information on polymer characteristics. Knowledge of polydispersity is also important because of its influence on the physical properties of polymers [1–3].

At present there is growing interest in tailor-made polymers as new polymeric materials, however, their properties cannot really be interpreted using average quantities. It is difficult to characterize polymers which have heterogeneity in chemical composition and monomer arrangement along the polymer chain. This heterogeneity needs to be taken into account along with the molecular weight of the polymer. The molecular-weight distribution (MWD) of a polymer is well known to be a function of the mechanism of formation, and it has often been stated that detailed knowledge of the molecular-weight distribution can provide valuable insight into the mode of the reaction [4]. Fractionation becomes important and in fact indispensible for characterization.

Polymer separation is based on molecular size differences. Various liquid chromatography techniques are employed to characterize the polymers. To fractionate polymers, several chromatographic techniques are commonly used with different separate mechanisms. Chromatography methods are

DOI: 10.5643/9781606502440/ch2

very sensitive to molecular characteristics. Separation according to molecular weight and composition should be carried out prior to characterization of the polymer.

2.1. Chromatographic Methods

Separation and characterization of polymers are performed using various chromatographic techniques. In the past, various methods were used for the determination of polymer–polymer interaction parameters. The methods were based on ternary systems consisting two polymers and a common solvent. Chromatographic methods provide for the comprehensive characterization of complex macromolecules with multiple distributions.

Despite the numerous advantages of chromatographic methods, there are also some difficulties, such as dissolving the polymer in solvent, which represents interaction of a component of the mixed solvent with a polymer chain. The functional group or hetero atoms along the polymer chain can form bonds with different polar components of a mixed solvent. Bond formation leads to an increase of the hydrodynamic radius of the polymer coil during its dilution in mixed solvent. Solvent dissolution can also affect conformational transitions of the polymer chain [5].

2.1.1. Pumps

Three basic pump designs probably account for most of the commercially available metering pumps.

1. *Peristaltic pumps* are used for low-pressure applications, that is, for flow rates at or near those of gravity feed conditions. These are not positive-displacement systems. The feed rate for a given motor speed generally is not reproducible on a day-to-day basis.
2. *Syringe pumps* are used from quantities of a few milliliters to several liters. They are used at up to several thousand pounds pressure. The pumps are pulseless. They can be used with flow- and pressure-sensitive detectors without pulsation damping systems. Syringe pumps are not useful for recycling systems, however.
3. *Reciprocating piston pumps* are used for operations at up to several thousand pounds pressure. They can be readily used for conventional as well as recycling chromatography.

2.1.2. Columns

Many column designs are available, of both metal and glass, of fixed length and adjustable length. Variable-length columns often include a plunger that passes through the end fitting of the column, which can be moved in or out to obtain a given packing length inside the column. A column kit is available which includes a number of end fittings together with precision-bore tubing for assembling a variety of columns. Packed columns are available with complete instrumentation.

Extremely high resolution is reportedly attainable through the use of very narrow columns packed with fine materials. The pressures often required to achieve practical flow rates with narrow columns, of course, makes instrument design considerations more stringent (high-pressure pumps, low-volume pumping and detectors, safety considerations, etc.), and this must be kept in mind when considering their use. Retention on any column type is directly proportional to the phase ratio β, where β = (void volume)/(stationary-phase volume) of a column [6].

2.1.3. Column Substrates

Choice of a column substrate for a given fractionation depends primarily on the restrictions imposed by the solute/solvent combination. The column packing must be wettable in the system of interest and should not be degraded by it. If an elevated temperature of operation is indicated (to dissolve the sample, or to improve molecular diffusion), the column substrate must be stable at that temperature.

Many new column substrates have become commercially available [7]. They can be classified as being rigid, semirigid, or soft. The rigid substrates, characterized by a fixed, rather uniform, pore volume, high column permeability, and ability to be "wet" by water and organic solvents, include porous silica and porous glass. These substances are not affected by most solvents and therefore permit separations to be achieved in solvents that degrade many organic substrates.

2.1.4. Detectors

Modern detector technologies open up new ways to investigate various properties with high sensitivity even in the low concentration ranges used in chromatography. Polymer separation techniques are combined with an appropriate detector(s) for the online or offline determination of various molecular characteristics.

Many kinds of liquid chromatographic detectors have been used and/or are available commercially. These are based on a variety of physical properties of molecules in solution. Some rely on the presence of a specific chemical functionality. Others do not, and are more universal in their scope of applicability. Often multiple detectors are used, one being nonspecific (e.g., a differential refractometer) to indicate the elution of all sample components, another being a specific detector (e.g., a UV photometer) to detect the presence of a specific chemical type in the eluate. With a suitably selected dual detector system, information can be obtained concerning both the molecular-weight distribution and the relationship between the MWD and the chemical composition of copolymers.

Liquid chromatographic detectors include differential refractometers, UV/visible refractometers, flame ionization detectors, heat-of-absorption detectors, electrical conductivity detectors, infrared detectors, and gravimetric detectors.

- Differential refractometers are based on refractive-index measurements and are extremely sensitive. They are almost universally applicable, as they do not rely on the presence of specific functional groups. Many polymeric systems exhibit a constant refractive index over a wide molecular-weight range, and it is often possible to relate the detector response to the sample quantity eluting from the column.

- UV and visible photometers are limited to applications in which the solute of interest absorbs radiation of the wavelength employed, whereas the solvent does not. These photometers, however, offer extreme sensitivity with good temperature and flow stability for specific solutes with high extinction coefficients.

- Flame ionization is allowed to fall on and wet a moving metallic wire, chain, or band. It then enters an oven maintained at a temperature high enough to completely evaporate solvent from the moving element, but without disturbing the residual sample. This residual material is then either passed directly through a flame ionization detector where it is pyrolyzed and ionized, or it is first pyrolyzed in a high-temperature furnace and then the pyrolyzate is blown into the flame ionization detector. Consequently, the currently available flame ionization detectors have not gained wide acceptance for liquid chromatography. Also, these detectors are limited to those applications where solvent can be removed from the moving element without removing any of the solute (sample components).

- Heat-of-absorption detectors consist of a microcolumn in which a temperature-sensing element is imbedded. The heat of adsorption and desorption is detected and measured as a change in temperature of the absorbent column. These detectors are very sensitive and useful for qualitative detection, but they require frequent calibration.
- Electrical conductivity detectors consist of a pair of metallic electrodes in a microcell together with a Wheatstone bridge measuring circuit. Applications are limited to systems that conduct electricity.
- Infrared detection has been limited to samples containing functional groups that exhibit strong absorption bands within the wavelength range of the instrument. Sensitivity is relatively poor except in cases where an extremely strong absorption hand can be monitored (e.g., carbonyl, C–H, etc.) [8,9].
- Gravimetric detectors involving direct automatic continuous weighing of the chromatographic effluent (to yield density), or automatic evaporation of solvent and weighing of the resultant residue from discrete microfractions, would seem to offer a most valuable detector for liquid chromatography based on the use of an electrobalance as the sensor, but it does not appear that a commercial instrument will be forthcoming in the foreseeable future [10].

Polymer separation techniques are combined with an appropriate detector(s) for the online or offline determination of various molecular characteristics. The most commonly utilized detectors with field-flow fractionation (FFF), discussed later, are multi-angle light-scattering (MALS), differential refractive index (dRI), ultraviolet and visible (UV/vis), differential viscometry, nuclear magnetic resonance (NMR), and Fourier-transform infrared (FTIR) detectors. MALS, dRI, UV/vis, and viscometry detectors offer the advantage of online detection. Without the separation step, the detection methods listed above provide average values representative of the entire sample population. Matrix-assisted laser desorption/ionization time-of-flight mass spectrometry (MALDI-TOF-MS) has also become a workhorse for macromolecular analyses [11–13].

2.1.5. Efficiency

In general, maximum efficiency can be realized with systems containing a minimum of dead volume with narrow-bore columns containing finely divided packing (substrate) and narrow particle size distribution. Small molecules are retained longer due to their ability to diffuse in the pores of the stationary phase. Large molecules elute first, and normally they are excluded from the pores. Separation and characterization of heterogeneous polymers (polymers with more than one type of distribution) is highly desired.

2.2.   Liquid Chromatography

Liquid chromatography (LC) has become one of the most important tools worldwide for separation, identification, and quantification tasks in analytical laboratories [14–16]. Liquid chromatography is a powerful instrument for polymer characterization [17,18]. Polymer properties such as molar mass distribution and chemical composition [19], as well as polymer–solvent interactions, can be assessed by LC [20]. Figure 2.1 shows a simple diagram of a liquid chromatography setup.

In liquid chromatography at critical conditions, entropic and enthalpic interactions are balanced through the selection of solvent and temperature. The separation which results is governed by small differences in the chemistry of the components [21]. It is used for the characterization of copolymers, or functional-type distributions. It is highly sensitive to temperature and solvent composition fluctuations, which can lead to peak splitting, peak

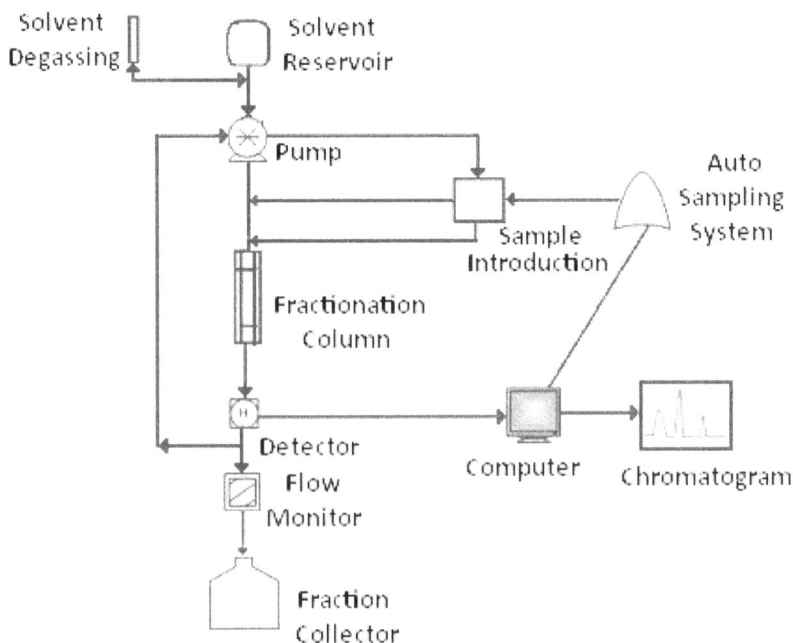

Figure 2.1.   Simple diagram of a liquid chromatography setup.

broadening, analyte loss due to adsorption to the stationary phase, and limited reproducibility. Chromatographic techniques with different separation mechanisms are commonly used to fractionate polymers.

Liquid chromatography at limiting conditions (LCLC) belongs to the so-called barrier polymer high-performance liquid chromatography (HPLC) approaches. LCLC is less sensitive to mobile-phase composition or temperature changes than LCCC (liquid chromatography with critical conditions). It also has a broader analyte molecular-weight operating range, and accommodates column overloading [22].

Preferential solvent dissolution plays an important role in thermally induced polymer diffusion [23], radical polymerization [24], and other processes, so it is important to know as much as possible about the separation process and be able to estimate molar mass well. Interpretation of the chromatographic results should be considered in light of this knowledge, because it leads to determination of the qualitative effect and a quantitative estimation of the phenomenon of preferential solvent dissolution analysis [25–28].

2.2.1. Requirements for Liquid Chromatography

- Separation—resolution R, or peak capacity PC
- Sensitivity—peak volume
- Speed—equivalent to gradient time
- Material recovery, band shape
- Use of convenient equipment

2.2.2. Separation Variables

- Sample (molecular weight, chemical structure)
- Equipment (flow-rate range F, extra-column volume u, detector time constant t, etc.)
- Column configuration (dimensions L and d, particle diameter dp and pore size, composition of bonded phase, plate number N and permeability, etc.)
- Experimental conditions (apart from mobile phase) (gradient time to flow rate F, column length L, gradient range dq)
- Mobile-phase composition (buffer, pH, choice of organic solvent, ion pairing, etc.)
- Use of denaturing conditions (temperature, addition of chemical, etc.)

2.3.   High-Performance Liquid Chromatography

High-performance liquid chromatography (HPLC) has become one of the most important tools worldwide for separation, identification, and quantification tasks in analytical laboratories [14–16]. As noted earlier, HPLC at limiting conditions belongs to the so-called barrier polymer approaches. Liquid chromatography at limiting conditions is less sensitive to mobile-phase composition or temperature changes than LC at critical condition. HPLC is less sensitive to mobile-phase composition or temperature changes. HPLC can handle a broader molecular-weight range of polymers and accommodates column overloading. Figure 2.2 shows schematics of two types of HPLC systems.
   Advantages of HPLC include the following:

- Absorbent must be cleaned in situ.
- Columns can be cleaned and reused.
- No specialized sampling equipment is needed.
- Sample application can be automated.
- Results are repetitive, with automatic sample injection on the same column.
- No special precautions are needed.
- Detection is automatic and nondestructive.
- This process is automatic and involves no personnel time.
- Fractions are free from contamination, because of improved efficiency and reduced sample handling.
- Recovery is typically around 95–100%.
- It has a broad analyte molecular-weight operating range.
- It can handle a wide range of elution compositions [22].

2.4.   Gel Permeation Chromatography

Gel permeation chromatography (GPC), sometimes called size-exclusion chromatography (SEC), is a relatively new technique for fractionating polymers, but it has become the premier characterization method for homopolymers, condensation polymers, and strictly alternating copolymers. Today, GPC is one of the most powerful instrumental tools available to polymer scientists. It has been used extensively to characterize high-molecular-mass polymers and oligomers of various polymer classes [29]. It also permits determination of the molecular-weight distribution of polymers [30].

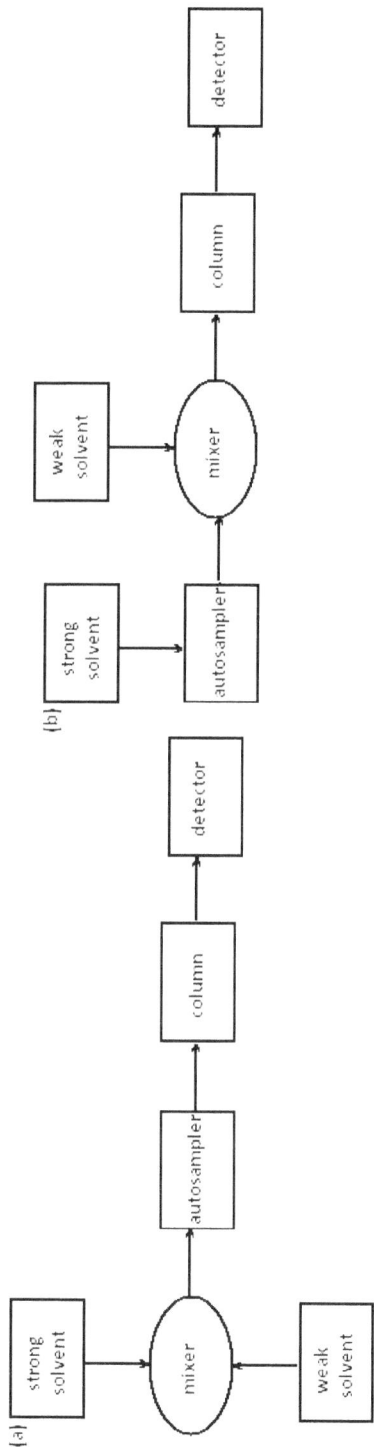

Figure 2.2. Schematics of HPLC systems in which the mobile phases are mixed (a) in front of the autosampler (conventional setup) and (b) after the autosampler. [Reprinted from E. Reingruber, F. Bedani, W. Buchberger, and P. Schoenmakers, *J. Chromatogr. A* 1217, 6595–6598(2010). with permission from Elsevier Science Ltd. Copyright © 2010 Elsevier B.V. All rights reserved.]

GPC is a chromatographic method in which polymer solution as liquid phase percolates through a porous gel as stationary bed [31]. Based on permeation in the gel column, solute molecules are separated into different internal volumes of molecules of different sizes over an extended range.

2.4.1. Basics of GPC

GPC is a special type of liquid–liquid partitioning in which the stationary phase is the solvent contained within the gel pores and the moving phase is the solvent outside the column substrate particles. GPC is based on the depth to which each molecular species is able to diffuse into the gel network. An entropically controlled separation occurs according to the hydrodynamic volumes or size of the molecules [32].

GPC operates on the principle that small molecules in a solution will diffuse into the pores of the column packing material and thus will elute more slowly. The larger molecules pass through the column more quickly, and are eluted first. Retention time is inversely proportional to molecule size. The eluted molecules are detected by one of several methods. Generally, detectors are either concentration-sensitive, such as a refractive-index detector, or molar mass–sensitive, such as a viscosity detector [33].

The separation of polymer molecules by GPC is based on the differences in their "effective size" in solution (effective size is closely related to molecular weight) [34]. The basis of the separations attained by these columns rests in the fact that access to the interior of the bed material bead is governed by the size of the solute molecule. Thus, in its simplest form, the column operates by diffusional partition of a solute between an interior stationary phase and a mobile exterior phase without the intervention of specific adsorption of the solute to the bed matrix [35–37] and sedimentation coefficient [38–40].

The capability of sorting molecules on the basis of molecular size and shape has extended the scope of GPC with the development of more universally applicable column substrates. Its scope includes both very high- and very low-molecular-weight polymers, and it can be used with a variety of solvents, both aqueous and organic, and for industrial-scale separations and purifications.

2.4.2. Separation

In gel permeation chromatography, a dilute polymer solution is injected into a solvent stream, which then flows under pressure through a chromatography column filled with porous gel packing, typically composed of silica beads and/or a polymeric gel. GPC separates the polymer molecules according to the size of the polymer chain in the elution solvent. The separation depends on the ability of the solute molecules to penetrate into the gel. It utilizes the partition equilibrium of polymer chains between common solvent phases located at the interstitial space and the pores of the column packing materials, typically in the form of uniform-size porous beads. Given the relationship between the chain size and the molecular weight of polymers, especially linear and chemically homogeneous polymers, GPC retention is well correlated with molecular weight.

The amount of internal pore volume accessible to a given polymer molecular species depends on its size. The pore size of the gel particles may vary from small to very large. The pore size in simple cases correlates with molecular weight, since larger species permeate less into the internal pore structure of the gel. Larger species are eluted first [41].

For fractionation of highly polydisperse samples, different columns can be used in series. Control of the separation is exerted through control of the porosity of the gel. Thus the term *gel permeation* is preferred to gel filtration [42] or molecular-sieve chromatography [43]. The volume elution curves, called chromatograms, represent the fractionation.

2.4.3. Mechanism

In GPC separation, the operation dominates under a given set of operating parameters. There are mechanisms to describe GPC as a liquid–liquid partition, steric exclusion, and restricted diffusion [44]. In liquid–liquid partition chromatography, the solute molecules distribute themselves between two liquid phases—the liquid contained in the gel pores and the liquid outside the gel, i.e., in the interstitial volume.

In the steric exclusion mechanism, it is assumed that different fractions of the total pore volume are accessible to different pore sizes. The process is expected to be flow rate–insensitive (not diffusion-controlled) over a relatively wide range of linear velocities. The effects of steric exclusion should be most evident where a major portion of the solute molecules are larger than many of the gel pores.

The small molecules diffuse most rapidly into the gel pores, so the probability of larger molecules diffusing into the gel pores is reduced. The larger molecules move until they find unoccupied pores. The net result is enhancement of the separation [45].

The molar volume of the solute and the pore size distribution of the gel accounts for separation based on the existence of different chain confirmations within either spherical, cylindrical, or slab-shaped pores [46–49]. The behavior of random-coil molecules in uniform pores is related to the probability that the root-mean-square end-to-end distance of the molecules will be less than the average pore radius[50].

If the restricted diffusion mechanism is dominant, the process is assumed to be diffusion-controlled to a significant degree, i.e., there is no diffusional equilibrium. Observed retention volume should be affected by changes in flow rate. The shapes of broad chromatographic bands observed for polydisperse mixtures should also be flow rate–dependent. The absence of diffusional equilibrium should be most pronounced at very high linear velocities.

There are five basic forces that one would expect to modify the elution volume of a compound in GPC:

1. Changes in the width of the network openings in the beads. With the same solvent this remains constant, but it changes with the solvent as a function of gel swelling [51].

2. Solvent–solute association. For example, the hydrogen bonding of OH compounds to ethers is well known [52–55]. Complete bonding of the solvent and solute often occurs, and then the size of the solute molecule is increased by an amount equal to the size of the solvent.

3. Dimerization, etc. For example, acetic acid and other organic acids in CCl4 form a dimer, as established by NMR and other measurements [56,57]. However, a dimer is not formed in water, where the hydrogen bonding of the acid's OH to the oxygen of the water predominates. In GPC, an aggregation of molecules will look two or more times as long as the normal chain length.

4. Intramolecular bonding. This is the formation of a ring structure when the groups that one might expect to form a solvent–solute association are located within the same molecule. Reportedly, an ether solvent such as dioxane will break up intermolecular bonding but not intramolecular bonding [58].

5. Adsorption onto the gel surface or into it. Solute–solvent interaction should be strong enough to suppress any tendency for the solute to be adsorbed to the gel. Since the gels used do not exhibit any strong electron-accepting or -donating groups, any active hydrogen, or any strong dipoles, this should be easy to achieve. If it occurs, adsorption is detected by tailing and/or elution later than would normally occur [58–60].

2.4.4.   Packing Materials

Gel permeation chromatography is largely controlled by the packing materials used in the column. The main problem with GPC is to find appropriate packing to obtain high selectivity and efficiency. The packing material should have the following properties:

1. It should be available as rigid beads with reproducibility and allow dense packing of the column.
2. Particles should be small, in the 5– to 10-μm size range, to minimize peak broadening in the column.
3. High porosity of the beads is desired to obtain a large intraparticle volume for fractionation.
4. The packing should have suitable graduated pore sizes to suit the separation of polymers in a wide molecular range.
5. The surface should be chemically inert, to prevent chromatography at high temperature.
6. The packing should have sufficient thermal stability to permit chromatography at high temperature.

The GPC process is basically a liquid–solid chromatographic method in which the separation is carried out by passing a dilute polymer solution through a column containing a rigid, but highly porous, cross-linked polystyrene gel packing. The solute molecules permeate the gel according to their volume in solution [59–61]. Since the higher-molecular-size species have less pore volume to permeate, they are eluted first. As is true with all chromatographic processes, dispersion during flow through the packed column and associated tubing and detector causes a spread in retention times about the mean. This peak spreading is a major factor limiting the accuracy of GPC analyses of polydisperse polymers.

Chromatographic columns filled with grains of pore-controlled glass display elution spectra which, except for somewhat sharper resolution, are similar to those shown by the widely used hydrogel or organogel columns. In general, substances of large molecular size do not enter the gel phase but move only in the interstitial fluid. These compounds thus emerge in the effluent immediately after the passage of a quantity of liquid equal to the void volume of the column. Compounds of low molecular weight can pass into the gel grains and are consequently retarded on the column. The retention is determined by the molecular dimensions of the solute and by the network of the gel but may also depend on other factors.

2.4.5.   Instrumentation

The GPC instrument has twin pumps with voltage stabilization, thermo-statting on the refractometer optical bench, and incorporates a non-servo-system refractometer with smaller cells. Samples are made up as solutions in solvent drawn from the main supply. The solutions then are filtered prior to injection to remove particles of gel or dirt which would otherwise block the GPC column.

The recorder sensitivity is kept constant, and the column set is calibrated using materials with known molar volumes of solute. The column set used consists of a known plate count [62].

Calibration of the columns is carried out with a series of well-characterized polystyrenes and poly(propylene glycols). Sometimes the viscosity molecu-lar weights determined from the intrinsic viscosities are checked against vis-cosity molecular weights calculated from the GPC curves [63].

The basic elements of a GPC setup are a solvent pumping system, a sam-ple injector, columns with provision for thermostatting, and a detector to record the amount of solvent eluted versus the polymer concentration. In the

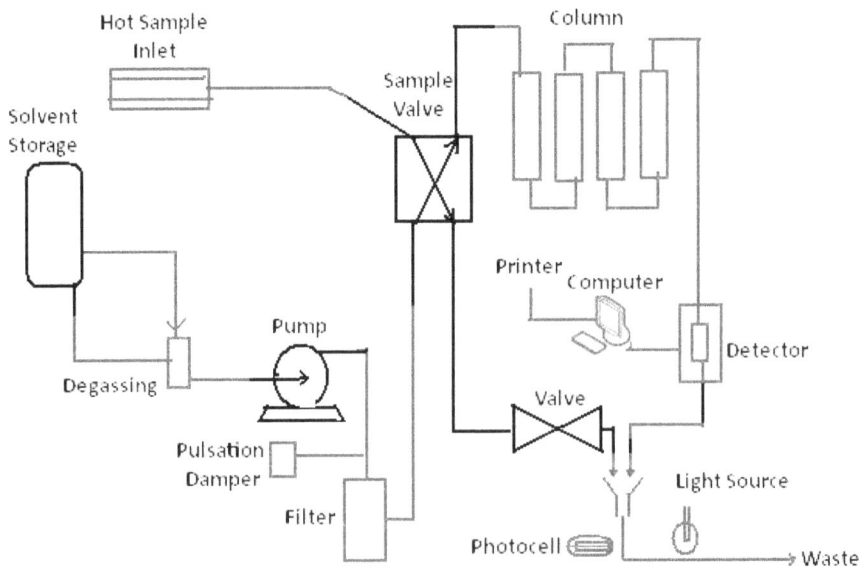

Figure 2.3.   Gel permeation chromatography setup.

setup shown in Figure 2.3, an infrared spectrophotometer is used as a detector. Another commonly employed type of detector is a differential refractometer. The end tubes are washed with acetone and dried with an air jet. This compartment is quickly emptied with the syringe.

2.4.6.   Sample Testing

A sample is prepared by dissolving a small amount of polymer in the solvent and filtering the solution to remove the undissolved impurities. The next step is to select the proper size columns, connect them, set the sensitivity setting on the detectors, and allow the instrument to equilibrate. A trial analysis is done by injecting the polymer solution into the instrument. The chromatogram is carefully analyzed. If the chromatogram shows all the desired information, the final analysis is carried out. If not, the operating parameters, such as column size, flow rate, and number of columns, are optimized. The trial step is repeated before proceeding to the final analysis.

During the final analysis, as the sample flows through the column, the molecules are separated according to size by a simple mechanical effect. The various molecular-weight species are separated by the difference in travel time through the column and pass through a detector in descending order of size. The detector measures the concentration of each molecular size and plots the molecular weight distribution of the sample on a strip chart.

2.4.7.   Calibration

According to the Einstein viscosity law, one can write

$$[\eta] = K(V/M) \tag{1}$$

where $[\eta]$ is the limiting viscosity index, V is the hydrodynamic volume of the particles, M is their molecular weight, and K is a constant.

Calibration means primarily establishing a relationship between the molecular weight of a monodisperse species (or very narrow fraction) and the emergence volume V corresponding to the peak maximum of the chromatogram, although the questions of peak broadening due to statistical effects and column imperfections require quantitative answers. A tentative approach to the first part of the problem can be based on the assumption that for a given set of experimental conditions, flexible-coil molecules of the same statistical dimensions will exhibit the same emergence volume, provided no specific interaction with the gel takes place. The following considerations support this assumption.

In the case of rigid molecules, it is agreed that the GPC process consists of a sequence of admissions to or exclusions from pores, depending on the relative physical sizes of the pores and molecules, with the result that smaller molecules are retarded to a greater extent in their passage along the column [63].

Flexible-coil molecules, on the other hand, are easily deformed, and we must consider partial penetrations accompanied by deformation as of particular importance in this case. Recent calculations by Meier [64] on the deformation of polymer coils between parallel plates suggest that the configurational part can be assumed to amount to at least 90%, even in thermodynamically good solvents.

GPC, a rapid new method for determining molecular-weight distributions of polymers [30], is a column chromatographic technique which sorts molecules according to their sizes in solution. Two methods are generally used to evaluate the performance of the columns and operating conditions: determination of the number of theoretical plates N using a low-molecular-weight, monodisperse (single-molecular-size) material and establishment of a calibration curve relating molecular weight to the peak elution volume V of the GPC curve [65]. In addition, there are formulas which can be used as tests for performance by calculating and predicting resolution and fractionation in GPC [66,67], as well as the maximum number of components resolvable by GPC [68].

Although the calibration curve is a useful measure of quality for the GPC operator, it depends specifically on polymer type and cannot be used easily to communicate quality between laboratories. Plate count, or number of theoretical plates, is a well-defined quantity used extensively in some areas of chromatography, but as presently determined it is not properly descriptive for GPC [67,68]. The very-low-molecular-weight, monodisperse material (such as acetone) is the last of the molecular sizes, except perhaps water and air, to be eluted from the column. The size-exclusion/diffusion mechanism [65,69] suggests that this material has taken the longest possible equilibrium route through the column and thus can be used to measure the maximum plate count affordable by the column. Unfortunately, the maximum plate count does not very well describe the real plate count available to much larger, more slowly diffusing polymer molecules, which see a smaller column volume. Because monodisperse polymer standards do not exist, polymers have not been used to determine plate count. The curve width W varies with dispersity, and it has been assumed that Eq. (2.1) [70] cannot be used when polydisperse materials are employed:

$$N = 16(v/W)^2 \qquad\qquad (2.1)$$

2.4.8. Column Packing

Columns to be used in a separation are calibrated by examining the elution volume of materials with very narrow molecular-weight distributions. A graph is prepared from the results showing the elution volume to sample peak versus the theoretical extended chain length of a molecule of the solute [71].

Porous glass is a material with excellent mechanical properties for gel permeation chromatography, and is capable of resolving polymers over a wide molecular-weight range. Columns may be packed easily and require no special precautions to exclude air. Physical properties such as pore-size distributions may be measured by conventional methods [72].

2.4.9. Chromatogram

In gel permeation chromatography, a polymer sample is injected either into tubing which runs into the column or directly into the column. After passing through the column, sample molecules encounter more tubing, followed by a detecting device to continuously record polymer concentration. The contributions to the elution curve of a polydisperse sample from the column and sources exterior to the column are formally evaluated [73].

When a polydispersed sample is being eluted through a GPC column, its chromatogram is broadened by two processes, a desirable process due to the difference in molecular size of the species and an undesirable process due to mixing in the longitudinal direction. The second broadening process impedes the resolution of the column. If the elution of a sample is allowed to proceed to some part of the column and then the direction of flow is reversed, the chromatogram of the eluent reflects only the effect of the second process [74].

The chromatogram of a monomeric compound appears not as a straight line but as a bell-shaped Gaussian curve, as shown in Figure 2.4. The peak position is such a case depends on the molecular weight of the compound. The area under the curve is proportional to the weight or concentration of the compound in solution. The width of the Gaussian curve depends on the resolution of the chromatographic column.

For a polydispersed sample the chromatogram is a composite of the Gaussian curves of all its components. The total area under the curve is still proportional to the concentration of the sample, but the height of the curve does not reflect the relative abundance of the components at the corresponding eluent volumes, as it depends also on the abundance of the neighboring components.

At the two extremities of the chromatogram there are curve portions representing components which do not even exist in the sample. It is clear that this overlapping and diffused pattern of the chromatogram must be accounted for in the calculation of the true molecular-weight distribution. A method to treat this problem has been described [75].

2.4.10.   Advantages

- GPC is the method of choice for determining the molecular weight distribution of a polymer.
- It has relatively low cost, simplicity, and the ability to provide accurate, reliable information in a very short time.
- It detects not only resin-based molecules such as polymers, oligomers, and monomers, but also most additives used in plastic compounds and even low-level impurities.
- Batch-to-batch uniformity can be checked quickly as a means of quality assurance.
- Differences in molecular-weight distributions, peak shapes, shifts, and tailing are readily observable.
- Comparisons of additives and other lower-molecular-weight species are straightforward.

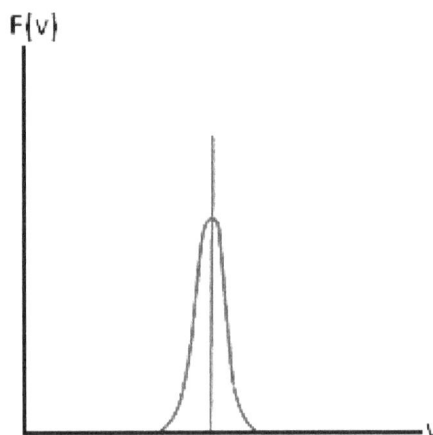

Figure 2.4.   Chromatogram of a monomeric compound.

2.4.11.   Disadvantages

- GPC is not an accurate method of determining molar mass, especially because the backbone or side-chain structures of analyte polymers differ significantly from those of the polymer size standards.
- GPC data can only be acquired for soluble components, not for insoluble conjugated polymers.
- Inherent shortcomings of the present organic column packings for GPC are solvent selectivity, temperature sensitivity, and mechanical instability [72].
- GPC is not efficient for separating polymer molecules according to molecular characteristics that do not have a simple correlation with chain size, such as chemical composition, chain architecture, and functionality.
- GPC is useful only for an estimation of molar mass; it is intrinsically empirical and cannot necessarily be used quantitatively nor accurately to determine the molar mass distribution [76,77].

2.5.   Field-Flow Fractionation

Field-flow fractionation (FFF) was first introduced by J. Calvin Giddings in 1966 [78] and is now in widespread use in research because of its potential and versatility. In FFF, the flow and thermal methods are used to characterize synthetic polymers. The basic principle and theory, used in combination with the effects of a laminar flow profile, lead to an exponential concentration profile of the analyte components as a result of their interactions with a physical field applied perpendicular to the flow of the carrier liquid.

Due to the need for highly detailed information about the molar mass, chemical composition, functionality, and molecular architecture of macromolecular materials, new analytical separation techniques with increased resolution, sensitivity, selectivity, and broader applications are constantly sought after. Field-flow fractionation (FFF) is a rapidly emerging technique that meets many of these needs. FFF can fractionate a wide range of analytes, including macromolecules, and colloids, and particulates suspended in both aqueous and organic solvent carriers [79–82].

There is a correspondence between elution volume and molecular weight; thus, chemically similar polymer standards of known molecular weight can be used for calibration. FFF can also be easily inserted into 2-D chromatography setups to provide more detailed information on copolymers [83].

2.5.1. Basics of FFF

The terminology used to describe different modes of operation is historically based and follows the development of FFF. The "normal" mode was the first observed mode of operation [84]. The term "steric" evolved later to describe a second separation mechanism observed when FFF was extended to particles >1 μm [85]. As the flow rate was increased to shorten the analysis time, it was observed that the micrometer-sized particles eluted significantly earlier than predicted by the steric-mode retention-time equation.

Field-flow fractionation is an elution-based chromatography. In this method, the separation is carried out in a single liquid phase. It is characterized by the use of an external field applied perpendicularly to the direction of sample flow through an empty, thin, ribbonlike channel (Figure 2.5).

Flow velocity increases from near zero at the channel walls to a maximum at the center of the channel (Figure 2.5). The perpendicularly applied force drives the sample toward the accumulation wall. A counteracting diffusive force develops due to the concentration buildup at the wall and drives the analyte back toward the center of the channel. When the forces balance, steady-state equilibrium is reached and an exponential analyte concentration profile is built up. Retention occurs when analytes reside in flow velocity zones slower than the average flow velocity of the carrier liquid passing through the channel. Separation occurs because different analytes reside in different flow velocity zones.

The normal mode of separation, in which diffusion plays an important role in controlling component distribution across the channel, is the most widely used mechanism [20,24]. A schematic illustration of the basic principle of

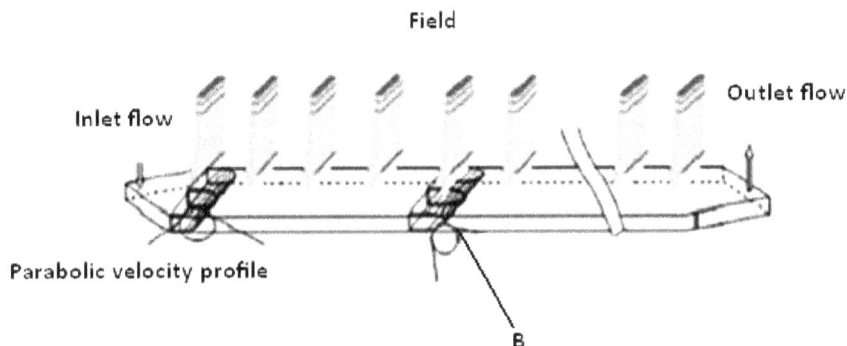

Figure 2.5.   Schematic representation of an FFF channel cut-out. [Reprinted with permission from F. A. Messaud and R. D. Sanderson, *Prog. Polym. Sci.* 34, 351–368 (2009). Copyright © 2009 Elsevier Ltd. All rights reserved.]

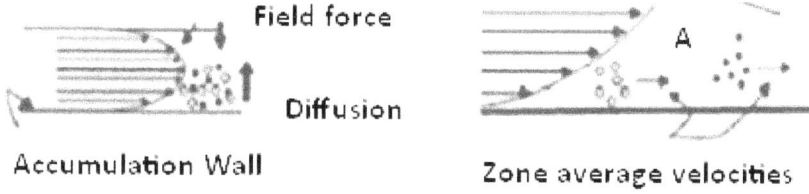

Field force

Diffusion

A

Accumulation Wall

Zone average velocities

Figure 2.6. Exploded view of the normal-mode separation mechanism of two components A and B (faster-diffusing B components are located at higher elevation in faster flow velocity streamlines and are thus eluted earlier than slower-diffusing A components). [Reprinted with permission from F. A. Messaud and R. D. Sanderson, *Prog. Polym. Sci.* 34, 351–368 (2009). Copyright © 2009 Elsevier Ltd. All rights reserved.]

normal-mode FFF separation is shown in Figure 2.6, and a typical resulting fractogram is shown in Figure 2.7.

The fractogram is a curve of detector response versus elution time (or elution volume). Analytes can be separated by different mechanisms (modes of operation) in FFF that arise from different opposing forces The mode of operation determines the elution order of analytes, along with other separation characteristics such as selectivity and resolution.

Three widely used modes that can be implemented in any FFF technique are normal, steric, and hyperlayer [80,84,87], as shown in Figure 2.8. The normal mode (based on Brownian motion of the analyte in the channel) is usually used for analyte sizes smaller than 1 μm. Smaller component populations accumulate in regions of faster streams of the parabolic velocity profile and elute earlier than larger components.

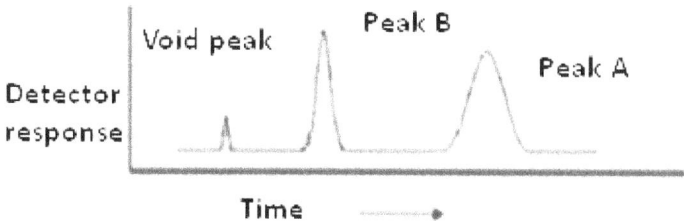

Void peak

Peak B

Peak A

Detector response

Time

Figure 2.7. A typical FFF fractogram. [Reprinted with permission from F. A. Messaud and R. D. Sanderson, *Prog. Polym. Sci.* 34, 351–368 (2009). Copyright © 2009 Elsevier Ltd. All rights reserved.]

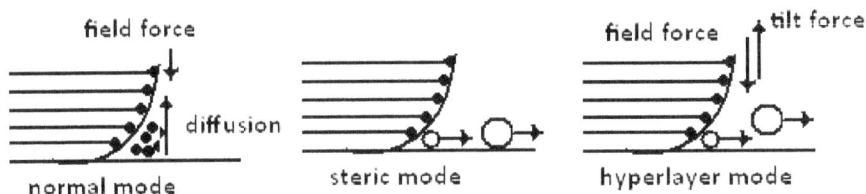

Figure 2.8. Schematic representation of different modes of operation that can occur in FFF. [Reprinted with permission from F. A. Messaud and R. D. Sanderson, *Prog. Polym. Sci.* 34, 351–368 (2009). Copyright © 2009 Elsevier Ltd. All rights reserved.]

2.5.2. Separation Mechanism

It became evident that these particles' centers of mass were located in flow streams well above one particle radius from the FFF accumulation wall [86–88]. This led to the introduction of the term "hyperlayer" to describe the formation of analyte layers above the channel wall as a result of two opposing forces [89]. While steric-mode separations are experimentally achievable, most separations involving micrometer-sized analytes are usually partially, if not entirely, in the hyperlayer mode.

Steric mode is applicable for components larger than ~1 μm, where diffusion becomes negligible and retention is governed by the distance of closest approach to the accumulation wall. Small particles can approach the accumulation wall more closely than large particles and thus the former's center of mass is in the slower-flowing streamlines.

The elution order in steric mode is from largest to smallest. Finally, lift or hyperlayer mode is one in which lift forces drive sample components to higher-velocity streams located more than one particle radius from the accumulation wall. These hydrodynamic lift forces occur when high flow velocities are used. The elution order is the same as in the steric mode. Most polymers are separated by the normal-mode mechanism because their dimensions are less than 1 μm [80–84].

2.5.3. Efficiency and Resolution

Due to the assumption of an exponential concentration profile of the analyte and of the parabolic flow profile used in the theoretical development of FFF retention theory, significant errors in retention parameters may arise under certain conditions [84]. FFF measurements also involve deviations between the experimental results and the expected theoretical behavior due to various effects, including zone broadening, overloading, solvent effects, analyte–wall interactions, analyte–analyte interactions, and nonuniformity of field strength. Some of these effects, e.g., zone broadening, are unavoidable and can only be corrected empirically. Many effects can be minimized by precise control and measurements of the different parameters such as flow rates, concentration, and temperature.

2.5.4. Advantages

- FFF overcomes some of the common limitations of traditional chromatographic techniques.
- It can overcome many of the limitations of current separation techniques, especially for ultrahigh-molecular-weight analytes and microgels.
- Fractionation of polymers can be simultaneously possible based on different physicochemical characteristics such as size and composition.
- There is no stationary phase, hence there are no sample breakthrough effects or sample loss due to adsorption to the stationary phase.
- FFF provides effective separation of microgel components simultaneously with solubilized polymer [90], with an upper limit of FFF extending to the 109-Da molecular-weight range and micrometer-size particles.
- There is minimal shear degradation [84,90,91].
- Materials separation can be accomplished with high resolution over a wide range from 1 to 100 nm [80,84].
- The process is physically simple and stable and can accommodate different types of samples.
- It is easy to adjust experimental conditions.
- Mild operations allow the analysis of fragile samples [82,92].
- The systems are physically simple and stable.
- There is minimum sample lose.
- It is easy to adjust or change experimental conditions to accommodate different kinds of samples.
- Operating conditions are gentle, and maintenance is easy.

2.6.   Super-Fluid Chromatography

Super-fluid chromatography (SFC) is used in the analysis of relatively non-polar polymers. Attempts to extend it to more polar solutes go back to at least 1969 [93]. SFC has become a popular tool to separate polymers from their oligomers, but polar solutes either do not elute or elute with poor peak shapes from packed columns, using pure carbon dioxide as the mobile phase. Using highly efficient and selective capillary columns, however, unique separations may be achieved. The solute range of interest is from phenol and aniline, the least polar, to polyfunctional aliphatic amines and polyfunctional acids at the most polar end. However, limited experimental data are available on simultaneous separations according to functionality and molar mass.

2.6.1.   Mobile Phase

It is important to select a mobile phase of binary and ternary mixtures of carbon dioxide with organic solvents and additives. These fluids are rarely used with capillary or open tubular columns for polar solutes involving packed-column separations. There is no significant difference between subcritical-fluid chromatography (subSFC or sSFC) and near-critical-fluid chromatography. The mobile phase enhances fluidity chromatography and SFC. Transitions between sSFC and near-critical-fluid chromatography are undetectable chromatographically, and the instrumentation used is identical.

SFC uses moderately polar fluids such as pure isopropanol, at high temperatures with non–silica-based, microporous polymer beads as the stationary phase. Pure carbon dioxide [94] is surprisingly polar and should dissolve relatively polar solutes [93] at much lower temperatures.

Carbon dioxide is similar to pentane, not isopropanol, yet it is widely believed to be more polar. For different reasons, carbon dioxide has become the fluid of choice for SFC. To have an effective chromatographic mobile phase, solutes must have appreciable solubility in a fluid. Polar functional groups are surprisingly soluble in carbon dioxide. More polyfunctional solutes tend to be much less soluble. Formic and acetic acids are miscible. Both phenol and aniline are soluble up to 3% [95,96]. However, it is difficult to elute from either a capillary or a packed column with pure carbon dioxide. Less soluble solutes are less likely to be eluted with carbon dioxide.

With its very low solubility, chromatography with carbon dioxide alone would be doubtful. There have been many attempts to separate many polar solutes using pure carbon dioxide. Carbon dioxide is used as fluid of choice due to its modest critical point, low cost, available purity and safety, ease of use, and lack of a viable alternative. The use of supercritical ammonia for the separation of polar molecules, including biopolymers, results in irreproducibility. Moreover, ammonia varies with its water content, and potential dangers are associated with its use.

2.6.2.   Stationary Phase

In SFC, the retention involves comparing on capillary and packed columns. The packed columns are 10–100 times more retentive than typical capillaries [97]. Both capillary and packed columns will produce different retention characteristics using same fluids. Packed columns require greater inherent retention capacity with modified mobile phases.

Changes in retention are attributed to changes in the density of the fluid. Polar modifiers dramatically change the retention of polar solutes. The changes in the density of binary fluids cause smaller shifts in retention [98].

2.6.3.   Pumps

SFC systems use syringe pumps, operated as pressure sources. Flow is passively controlled with a fixed restrictor mounted on the end of the column. The main problem with the syringe pump is the inability or at least great difficulty in controlling parameters such as flow rate, density, temperature, and composition. These are all independent of each other, and determine the effect of any one variable on retention. Such systems are ideal for controlling pressure, but not flow. A change in pressure results in all ill-defined change in flow [99]. Both the density and the amount of fluid change with pressure. The effect of density on retention is intricately linked with the effect of changes in the flow rate on retention. The pumps must be used as flow sources, with an independent device controlling pressure to generate precise compositions.

Modern packed-column instruments use multiple high pressures, reciprocating pumps, operated as flow sources. There is independent control of system pressure through the use of electronic back-pressure regulators. This allows accurate, reproducible composition programming while retaining flow, pressure, and temperature control. To control flow accurately, the pump must have a larger-than-expected compressibility compensation range.

2.6.4.   Pressure Control

The use of long columns resulted from a change in control philosophy. Earlier, the pump was used as the pressure controller. The column outlet pressure was not controlled. Long columns produced large pressure drops, and, at modest inlet pressures, the outlet pressure could drop to the point where several (subcritical) phases could exist. The coexistence of several phases destroys chromatographic separations and efficiency. By controlling the column outlet pressure, the pump becomes a flow source, not a pressure source. Consequently, the point in the system with the worst solvent strength becomes the control point. All other positions in the system have greater solvent strength. By controlling this point, problems associated with phase separations or solubility problems at uncontrolled outlet pressures are eliminated.

2.6.5. Ovens

For many years, it was widely assumed that the fluids used as supercritical mobile phases lost their interesting characteristics when they became subcritical. Consequently, users strenuously avoided subcritical conditions. There was often a debate about whether reports using binary fluids represented HPLC or SFC. It was implied that subcritical conditions were inherently, dramatically inferior (much lower optimum velocities, much higher pressure drops) to supercritical conditions. There was a tendency to work at relatively high temperatures to avoid possible phase separations. It is now widely appreciated that the defined state of the fluid is generally irrelevant. Viscosity, diffusion coefficients, density, and solvent strength are nearly identical for just-supercritical or just-subcritical fluids with the same composition. Phase separations almost never occur if the pressure remains high enough. Ovens used for SFC have traditionally been GC ovens. However, with packed columns, high temperatures tend to be less important, while lower, even subambient, temperatures tend to be more important. This has led to the use of different kinds of ovens. A maximum temperature of 150°C is not unreasonable. A minimum temperature of less than 0°C is desirable.

2.6.6. Advantages

- SFC is compatible with flame ionization detection (FID). FID provides universal, sensitive detection of carbon compounds, with uniform response factors, while allowing pressure or density programming.

2.6.7. Disadvantages

- Solute polarity appears to be related to the solubility of very polar compounds in the mobile phases used.
- Substances requiring aqueous conditions or aqueous, ionic buffers to dissolve are, at present, poor candidates for separation by SFC.

2.7. Gas Chromatography

Gas chromatography (GC) is well established as a simple and rapid means of establishing the elemental analysis of polymer materials. For volatile samples, GC is used for mixture separation. For nonvolatile or thermally labile materials, high-pressure liquid chromatography, or just liquid chromatography, is used.

Gas chromatography GC is based on the distribution of a compound between two phases. In gas–solid chromatography (GSC), the phases are gas and solid. The injected compound is carried by the gas through a column tilled with solid phase, and partitioning occurs via the sorption–desorption of the compound (probe) as it travels past the solid. Separation of two or more components injected simultaneously occurs as a result of differing affinities for the stationary phase. In gas–liquid chromatography (GLC), the stationary phase is a liquid coated on a solid support.

2.7.1. Basics of GC

Gas chromatographic methods can be used for making a wide range of physical measurements. The separation effectiveness of volatile organic materials in GC has led to its almost universal application in organic analysis. The most important application of GC on polymeric stationary phases has been in the field of polymer solution thermodynamics.

In conventional analytical GC, the stationary phase is of interest only as far as its ability to separate the injected compounds is concerned. The knowledge of the retention volume due to the dissolution of a substance makes it possible to calculate relevant thermodynamic characteristics of the solution process, namely, the partition coefficient, the activity coefficient, and the change in the excess partial molar thermodynamic functions of the solute in the given stationary phase [100–105].

2.7.2. Instrumentation

The elements of a simple gas chromatograph are shown in Figure 2.9. A continuous steady flow of carrier gas, usually helium or nitrogen, passes through some form of flow-control valve and into the chromatographic column at a known pressure. Provision is made for introducing samples of vapor into the carrier gas system upstream from the column, either by injection from a syringe through a septum, or via a sampling valve. In the chromatographic column, the sample is partitioned between the mobile gas phase and a stationary phase packed within the column. The stationary phase is often coated in a thin layer onto an inert support, in order to facilitate rapid interaction with the vapor. The sample vapor is analyzed in the effluent carrier gas with a suitable detector, such as a thermal conductivity cell or a flame ionization detector, and the carrier-gas volume flow rate is determined at the outlet using a soap-bubble flow meter.

The simplest gas chromatographic experiment consists of introducing a small, brief pulse of sample vapor into the carrier gas stream at the entrance of the column. Assuming the vapor does not interact with stationary phase but is simply carried through the column by the carrier gas flow, it will be eluted in a shorter time. The sample elution time will take longer when the vapor sample interacts with the stationary phase.

2.7.3. Advantages

- GC enables diffusion coefficients to be calculated from the variation of the chromatographic peak width with carrier-gas flow rate [106].

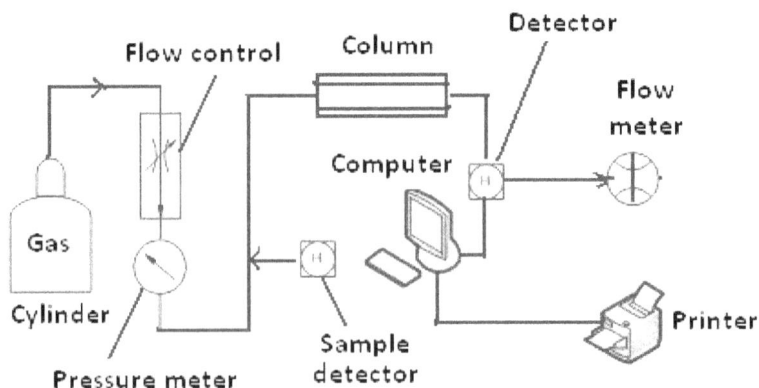

Figure 2.9. Simple gas chromatography setup.

- GC can be used in determining other polymer properties, such as glass transition temperatures, as reported by Smidrod and Guillet [107–109], melting temperatures [110], and the degree of crystallinity of the polymer [111,112].

- GC has received general recognition as an effective and simple technique for the rapid measurement of polymer solubility parameters [103,113] (by applying the Hildebrand-Scatchard theory) and polymer–solvent interactions in molten homopolymers [114,115].

2.7.4. Disadvantages

- Shapes of eluted peaks in GC on polymeric substrates are governed by several factors, one of the most important being the slow diffusion in the polymer phase [116].

2.8. Future Trends

Chromatography is a powerful tool for the study of polymers. The molecular-weight distributions of polymers are well known to be functions of the mechanism of formation, and it has often been stated that a detailed knowledge of the molecular-weight distribution can provide valuable insight into the mode of the reaction [117].

To improve features of LC methods, various combinations of coupled techniques with MALDI and mass spectrometry [118], multi-angle light scattering [119], Fourier-transform infrared spectroscopy (FTIR) [120], etc., are used for investigation of polymer systems. Attention has been toward the chemical point of view to use the technique in structural analysis.

References

1. H. V. Boeing, *Polyolefins: Structure and Properties,* pp. 93 and 132, Elsevier, Amsterdam (1966).
2. C. Jén-Yüan, *Determination of Molecular Weights of High Polymers,* p. 27, Israel Program for Scientific Translations, Jerusalem (1963).
3. F. Bueche, *Physical Properties of Polymers,* p. 222, Interscience, New York (1962).
4. F. W. Billmeyer, Jr., in J. Mitchell, Jr., and F. W. Billmeyer, Jr. (eds.), *Analysis and Fractionation of Polymers* (*J. Polym. Sci. C,* suppl.), p. 161, Interscience, New York (1965)

5. I. I. H. Park, *Polymer* 40, 2003 (1999).

6. T. A. Berger and J. F. Deye, *J. Chromatogr.* 594, 291 (1992).

7. J. Cazes, *J. Chem. Educ.* 43, A567, A625 (1966).

8. F. Rodriguez, R. A. Kulakowski, and O. K. Clark, *Ind. Eng. Chem. Prod. Res. Dev.* 5, 121 (1966).

9. J. H. Ross and M. E. Castro, *J. Polym. Sci. C* 21, 143 (1968).

10. J. L. Radawski and C. J. Williams, A gravimetric detector for liquid chromatography, Pittsburgh Conference on Analytical Chemistry and Applied Spectroscopy (March 1970).

11. R. Jagtap and H. Ambre, *Bull. Mater. Sci.* 28, 515 (2005).

12. H. Pasch and W. Schrepp, *MALDI-TOF Mass Spectrometry of Synthetic Polymers,* Springer-Verlag, Berlin (2003).

13. F. Basile, G. E. Kassalainen, and S. K. R. Williams, *Anal. Chem.* 77, 3008 (2005).

14. G. W. Somsen, C. Gooijer, and U. A. Th. Brinkman, *J. Chromatogr. A.* 856, 213 (1999).

15. M. Gallignani and M. R. Brunetto, *Talanta* 64, 1127 (2004).

16. M. Kölhed, B. Lendl, and B. Karlberg, *Analyst* (Cambr.) 128, 2 (2003).

17. B. G. Howard, B. E. Barry, and J. Christian, *Anal. Chem.* 70, 251R (1998).

18. B. Trathnigg, in R. A. Meyers (ed.), *Encyclopedia of Analytical Chemistry,* p. 8008, John Wiley, Chichester, U.K. (2000).

19. B. Trathnigg, *Prog. Polym. Sci.* 20, 615 (1995).

20. D. J. Winzor, *J. Biochem. Biophys. Meth.* 56, 15 (2003).

21. S. M. Graef, A. J. P. Van Zyl, R. D. Sanderson, B. Klumperman, and H. Pasch, *J. Appl. Polym. Sci.* 88, 2530 (2003).

22. D. Berek, *Macromol. Symp.* 231, 134 (2006).

23. A. C. van Asten, W. Th. Kok, R. Tijssen, and H. Poppe, *J. Polym. Sci. B* 34, 283 (1996).

24. Yu. D. Semchikov, L. A. Smirnova, N. A. Kopylova, and V. V. Izvolenskii, *Eur. Polym. J.* 32, 1213 (1996).

25. D. Berek, T. Bleha, and Z. Pevni, *Polym. Lett. Ed.* 14, 323 (1976).

26. T. Spychaj, D. Lath, and D. Berek, *Polymer* 20, 1108 (1979).

27. I. Katime and C. Strazielle, *Makromol. Chem.* 178, 2295 (1977).

28. L. Gargallo and R. Deodato, *Adv. Colloid Interface Sci.* 21, 1 (1984).

29. R. G. Beri, L. S. Hacche, and C. F. Martin, in J. K. Swadesh (ed.), *HPLC Practical Industrial Applications,* chap. 6, CRC Press, Boca Raton, FL (1997).

30. J. C. Moore, *J. Polym. Sci. A* 2, 835 (1964).

31. A. I. M. Keulemans, *Gas Chromatography,* p. 2, Reinhold, New York (1959).

32. H. G. Barth, B. E. Boyes, and C. Jackson, *Anal. Chem.* 70, 251R (1998).

33. L. H. Sperling, *Introduction to Physical Polymer Science,* 3rd ed., Wiley-Interscience, Toronto (2001).

34. L. H. Tung, *J. Appl. Polym. Sci.* 13, 775 (1969).

35. P. Andrews, *Nature* 196, 36 (1962).

36. S. Hjertén, *Arch. Biochem. Biophys.* 99, 466 (1962).

37. J. R. Whitaker, *Anal. Chem.* 35, 1950 (1963).

38. P. Flodin and J. Killander, *Biochim. Biophys. Acta* 63, 403 (1962).

39. S. D. Roskes and T. E. Thompson, *Clin. Chim. Acta* 8, 489 (1963).
40. R. L. Steere and G. K. Ackers, *Nature* 196, 475 (1962).
41. J. C. Moore, Pittsburgh Conference on Analytical Chemistry and Applied Spectroscopy (March 1963).
42. J. Porath and P. Flodin, *Nature* 183, 1657 (1959).
43. D. M. W. Anderson and J. F. Stoddart, *Anal. Chim. Acta* 34, 401 (1966).
44. K. H. Altgelt, in J. C. Giddings and R. A. Keller (eds.), *Advances in Chromatography,* vol. 7, pp. 3–16, Marcel Dekker, New York (1968).
45. K. H. Altgelt, Preprints, Div. Petroleum Chem., Am. Chem. Soc., 15 (No. 2), A-115 (Feb. 1970), a paper presented at the Symposium on Gel Permeation Chromatography, Houston, TX, Feb. 22–27, 1970.
46. N. V. B. Marseden, *Ann. N.Y. Acad. Sci.* 125, 428 (1965).
47. A. J. DeVries, M. LePage, R. Beau, and C. L. Guillemin, *Anal. Chem.* 39, 935 (1967).
48. M. Le Page, R. Beau, and A. J. de Vries, *J. Polym. Sci. C* 21, 119 (1968).
49. E. F. Casasa, *J. Polym. Sci. B* 5, 773 (1967).
50. J. B. Carmichael, *J. Polym. Sci. A* 2, 517 (1968).
51. R. H. Boundy and R. F. Boyer, *Styrene, Its Polymers, Copolymers and Derivatives,* p. 426, Reinhold, New York (1952).
52. T. S. Glasstone, *Textbook of Physical Chemistry,* pp. 111–114, Van Nostrand, New York (1940).
53. L. Pauling, *The Nature of the Chemical Bond,* Cornell University Press, Ithaca, NY (1940).
54. E. D. Becker, *Spectrochim. Acta* 17, 436 (1961).
55. G. C. Pimentel and A. L. McClellan, *The Hydrogen Bond,* p. 150, W. H. Freeman, San Francisco (1960).
56. T. S. Glasstone, *Textbook of Physical Chemistry,* p. 500, Van Nostrand, New York (1940).
57. G. C. Pimentel and A. L. McClellan, *The Hydrogen Bond,* W. H. Freeman, San Francisco, pp. 1–189 and pp. 193–201 (1960).
58. B. Ackermark, *Acta Chim. Scand.* 15, 985 (1961).
59. J. C. Moore and M. C. Arrington, The separation mechanism of gel permeation chromatography: Experiments with porous glass column packing materials, presented at the 3rd International Seminar on Gel Permeation Chromatography, Geneva, Switzerland (May 1966).
60. H. Benoit, Z. Grubisic, P. Rempp, D. Decker, and J. G. Zilliox, Gel permeation chromatography on linear and branched polystyrenes of known structure, presented at the 3rd International Seminar on Gel Permeation Chromatography, Geneva, Swizerland (May 1966).
61. W. B. Smith and A. Kollmansberger, *J. Phys. Chem.* 69, 4157 (1965).
62. W. B. Smith, J. A. May, and C. W. Kim, *J. Polym. Sci. A-2* 4, 365 (1966).
63. D. J. Meier, *J. Phys. Chem.* 7l, 1861 (1967).
64. J. C. Moore, *J. Polym. Sci. A* 2, 835 (1964).
65. D. D. Bly, paper presented at 154th American Chemical Society Meeting, Chicago, *Polymer Preprints* 1234 (1967).
66. D. D. Bly, *J. Polym. Sci. C* 21, 13 (1968).

67. J. C. Giddings, *Anal. Chem.* 39, 1027 (1967).

68. W. W. Yau and C. P. Malone, *J. Polym. Sci. B* 5, 663 (1967).

69. H. Purnell, *Gas Chromatography,* John Wiley, New York (1962).

70. G. D. Edwards, *J. Appl. Polym. Sci.* 9, 3845 (1965).

71. M. J. R. Cantow and J. F. Johnson, *J. Appl. Sci.* II, 1851 (1967).

72. H. Vink, *Makromol. Chem.* 116, 241 (1968).

73. L. H. Tung, J. C. Moore, and G. W. Knight, *J. Appl. Polym. Sci.* 10, 1361 (1966).

74. L. H. Tung, *J. Appl. Polym. Sci.* 10, 375 (1966).

75. W. Goetzinger, L. Kotler, E. Carrilho, M. C. Ruiz-Martinez, O. Salas-Solano, and B. L. Karger, *Electrophoresis* 1998, 19, 242–248.

76. F. C. Li, *Chromatogr. Sci. Ser.* 69, 249 (1995).

77. J. C. Giddings, *J. Sep. Sci.* 1, 123 (1966).

78. M. N. Myers, *J. Micro Column Sep.* 9, 15 (1997).

79. J. C. Giddings, *Science* 260, 1456 (1993).

80. S. K. R. Williams and D. Lee, *J. Sep. Sci.* 29, 1720 (2006).

81. S. K. R. Williams and M. A. Benincasa, in R. A. Meyers (ed.), *Encyclopedia of Analytical Chemistry: Instrumentation and Applications,* pp. 7582–7608, John Wiley, Chichester, U.K. (2000).

82. A. C. van Asten, R. J. van Dam, W. Th. Kok, R. Tijssen, and H. Poppe, *J. Chromatogr. A* 703, 245 (1995).

83. M. E. Schimpf, K. D. Caldwell, and J. C. Giddings (eds.), *Field-Flow Fractionation Handbook,* John Wiley, New York (2000).

84. J. C. Giddings and M. N. Myers, *Sep. Sci. Technol.* 13, 637 (1978).

85. F. A. Messaud and R. D. Sanderson, *Prog. Polym. Sci.* 34, 351 (2009).

86. J. C. Giddings, *Science* 260, 1456 (1993).

87. J. Chmelik, *J. Chromatogr. A* 845, 285 (1999).

88. K. D. Caldwell, T. T. Nguyen, M. N. Myers, and J. C. Giddings, *Sep. Sci. Technol.* 14, 935 (1979).

89. J. C. Giddings, *Sep. Sci. Technol.* 18, 765 (1983).

90. D. Lee and S. K. R. Williams, Separation of ultrahigh molecular weight polymers and gels, 56th Pittsburgh Conference on Analytical Chemistry and Applied Spectroscopy (2005).

91. J. Janca, *Field-Flow Fractionation: Analysis of Macromolecules and Particles,* Chromatographic Science Series, vol. 39, Marcel Dekker, New York (1988).

92. H. Lee, S. K. R. Williams, K. L. Wahl, and N. B. Valentine, *Anal. Chem.* 75, 2746 (2003).

93. S. T. Sie, L. E. A. Bleumer, and G. W. A. Rijnders, in C. L. A. Harboum and R. Stock (eds.), *Gas Chromatography,* pp. 235–251, Butterworths, London (1969).

94. J. C. Giddings, M. N. Meyers, L. McLaren, and R. A. Keller, *Science* 162, 67 (1968).

95. M. A. McHugh and V. J. Krukonis, *Supercritical Fluid Extraction,* p. 19, Butterworths, Boston, (1986).

96. A. W. Francis, *J. Phys. Chem.* 58, 1099 (1954).

97. T. A. Berger, in *Packed Column SFC,* chap. 4, RSC Chromatography Monograph Series, Royal Society of Chemistry, Cambridge, U.K. (1995).
98. T. A. Berger and J. F. Deye, *Anal. Chem.* 62, 1181 (1990).
99. T. A. Berger, *Anal. Chem.* 61, 356 (1989).
100. J. M. Braun and J. E. Guillet, *Adv. Polym. Sci.* 21, 108 (1976).
101. D. G. Gray, *Proc. Polym. Sci.* 5, 1 (1977).
102. R. J. Laub and R. L. Pecsok, *Physicochemical Applications of Gas Chromatography,* John Wiley, New York (1978).
103. J. R. Conder and C. L. Young, *Physicochemical Measurements of Gas Chromatography,* John Wiley, New York (1979).
104. J. E. G. Lipson and J. E. Guillet, in J. V. Dawkins (ed.), *Developments in Polymer Characterization,* vol. 3, chap. 2, Applied Science, London (1982).
105. G. J. Price, J. E. Guillet, and J. H. Purnell, *J. Chromatogr.* 369 273 (1986).
106. D. G. Gray and J. E. Guillet, *Macromolecules* 6, 223 (1973).
107. J. E. Guillet, in J. H. Purnell (ed.), *New Developments in Gas Chromatography,* p. 187, John Wiley, New York (1973).
108. O. Smirød and J. E. Guillet, *Macromolecules* 2, 272 (1969).
109. G. J. Courval and D. G. Gray, *Macromolecules* 8, 326 (1975).
110. J. M. Braun and J. E. Guillet, *Macromolecules* 10, 101 (1977).
111. J. E. Guillet and A. N. Stein, *Macromolecules* 3, 102 (1979).
112. G. DiPaola-Baranyi and J. E. Guillet, *Macromolecules* 11, 228 (1978).
113. K. Ito and J. E. Guillet, *Macromolecules* 12, 1163 (1969).
114. J. M. Braun and J. E. Guillet, *Adv. Polym. Sci.* 21, 107 (1976).
115. J. J. Van Deemter, F. J. Zuiderweg, and A. Klinkenberg, *Chem. Eng. Sci.* 5, 271 (1966).
116. F. W. Billmeyer, Jr., in J. Mitchell, Jr., and F. W. Billmeyer, Jr. (eds.), *Analysis and Fractionation of Polymers* (*J. Polym. Sci. C,* suppl.), p. 161, Interscience, New York (1965).
117. R. Murgasova and D. M. Hercules, *Anal. Chem.* 75, 3744 (2003).
118. P. J. Wyatt, in: J. Cazes (ed.), *The Encyclopedia of Chromatography,* p. 677, Marcel Dekker/Taylor & Francis Books, New York (2001).
119. K. Torabi, A. Karami, S. T. Balke, and T. C. Schunk, *J. Chromatogr. A* 910, 19 (2001).
120. A. Karami, S. T. Balke, and T. C. Schunk, *J. Chromatogr. A* 911, 27 (2001).

Chapter 3

Spectroscopic Techniques

Polymers have molecular weight, composition, and chain configuration. The molecular weight directly affects the physical properties of the polymer. To optimize the end-use properties, the polymer has to be adjusted. Characterization of the distributed molecular properties is the key to tailoring and optimizing a polymeric material. Hence an accurate knowledge of polydispersity is important because it influences the physical properties of polymers [1–3].

Applications of polymers can benefit from a clear understanding of molecular and structural parameters that control technological performance. Polymers are used to produce a great many versatile structures with a wide range of properties [4], and the molecular properties of the polymer directly affect the physical properties of the system. To optimize the properties of the end product, it is necessary to characterize and then adjust the molecular and physical properties in relationship to their intended end use. Instrumental methods are a key to tailoring and optimizing a polymer product.

Given the increasingly complex nature of polymeric materials, average properties are no longer adequate to characterize and elucidate the nature of the materials. Polymer testing methods are necessary to establish structure, property, and morphology relationships, which in turn are used to relate the polymerization mechanism and end-use performance. Molecular structure and properties have been widely studied in several scientific and technological disciplines. Polymers have been developed that have led to improved performance, new applications, and use of polymers to replace traditional materials.

DOI: 10.5643/9781606502440/ch3

Structure–property considerations are important in the development of new products at the molecular level to enhance and improve properties. Characterization of polymers is carried out by testing

- Molecular architecture and physical properties produced by polymerization
- Product processing so as to relate surface and bulk properties
- Morphologies related to application and end-use properties

An understanding of polymer synthesis, structure, and properties is thus always important. The physical properties of a polymer depend primarily up its morphology [5] before it is subjected to thermal and mechanical process during processing.

Important structural features include the actual arrangement of atoms in the polymeric "repeat unit" as well as the degree of order of long lengths of the chains. The characteristic properties are largely determined by the chemical nature of the polymer repeat unit. Spectroscopic techniques can be used to elucidate the chemical nature of the polymer, including the molecules in the bulk as well as molecules on surfaces and at interfaces. Most of these spectroscopic techniques involve measurements of band intensities.

3.1.   Fourier-Transform Infrared (FTIR) Spectroscopy

Infrared spectra with high spatial resolution have found particular favor in studies and analyses related to polymers and their end products [6,7]. IR spectroscopy has been utilized to study polymeric materials at the molecular level [8,9]. The main problem restricting the application of IR spectroscopy to polymeric samples has been that the atomic interactions in the IR portion of the spectrum are at relatively low energies.

Fourier-transform infrared (FTIR) spectroscopy is a popular and well-known technique for identification and measurement of compounds. It is an attractive choice owing to the unmatched wealth of molecular-level information contained in the infrared portion of the electromagnetic spectrum. FTIR spectroscopy requires only short times to carry out and produces results with good reproducibility and stability [10].

Further, FTIR overcomes the problems associated with low-transmittance heterogeneous samples [11]. FTIR techniques with transmission and reflection methods are well-established tools in many laboratories [12]. They are used routinely in "fingerprinting" contaminants. In polymer testing, particularly with FTIR, the use of specular reflectance measurements and the potential of infrared imaging which uses radiation from a synchrotron as the infrared source [13–18] have made it popular. The majority of polymeric materials are identified with the help of FTIR.

For FTIR, large databases of spectra stored in computers aid spectral interpretation when searching and comparing scanned spectra [19]. Spectral subtraction, multiple scanning, and various manipulation routines have further improved FTIR as a nondestructive analytical tool [11]. It also does not require any knowledge of mean molecular absorption.

3.1.1.   Principle

Infrared spectroscopy is based on the Beer-Lambert law. Therefore, the thickness and density of a sample are not needed. However, the peak absorbencies are proportional to the concentration of the sample.

The group concentration C with IR absorbance based on the Lambert-Beer law is

$$A = abC \qquad (3.1)$$

where

A = IR absorbance
a = absorbance coefficient
b = thickness of the sample
C = concentration of the evaluated group

The sample thickness varies from one sample to another. The internal standard method is employed to determine the concentration. The internal standard is a sample that contains the relevant functional group but is inert, such that its absorbance is related directly to its concentration. This method is widely used in many fields [20–22].

Molecular motions such as vibration, rotation, rotation/vibration, or lattice mode, or a combination, difference, or overtone of these vibrations result in a change in the molecule's dipole moment if a molecule absorbs radiation in the IR region of the electromagnetic spectrum.

Infrared light is applied to the polymer sample, resulting in the absorption of energy by individual molecules. The molecules then vibrate at excited energy states. This gives rise to absorption spectra that reflect the characteristic vibration bonds of the functional groups in the sample mole-cules [23]. Some of the infrared radiation is passed through (transmitted). The resulting spectrum represents molecular absorption and transmission, creating a molecular fingerprint of the sample.

The IR spectrum shows a fingerprint of the sample with absorption peaks. The peaks correspond to the frequencies of vibrations between the bonds of the atoms making up the material. Each material consists of a unique combination of atoms. Hence no two compounds produce the exact same IR spectrum.

Positive identification of polymeric materials is made by studying the size of the peaks in the spectrum. The amount of material present can also be. Characteristic changes in the polymeric sample are reflected in the recorded spectra [24].

3.1.2.   Instrumentation

Figure 3.1 shows the basic components of FTIR, for which the IR spectro-photometer contains an IR source, interferometer, sample holder, detector, and a signal and sample data processor.

The instrument includes a moving mirror, an optical bench, and a laser. The detector is connected through a source. The moving mirror, optical bench, and sample holder are also connected to the detector. From the detec-tor, the signal goes to an interferometer (Figure 3.2), which passes the signal to a computer connected to the IR instrument [25]. The computer converts the signal into spectra using data processing.

The IR source emits light in the IR region when electricity is passed through it. The beam splitter serves to split the incident IR light into two. The mirrors reflect the light waves in the direction of recombination of the waves at the beam splitter. The movable mirror is capable of moving away from and toward the beam splitter. The splitter splits the beam into two halves. One half of the split light goes to the stationary mirror and back to the beam split-ter. The other half of the split light is reflected onto the moving mirror. The

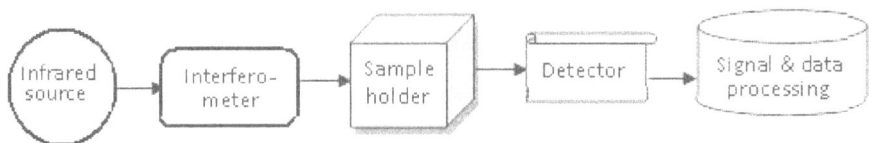

Figure 3.1.   Basic components of FTIR.

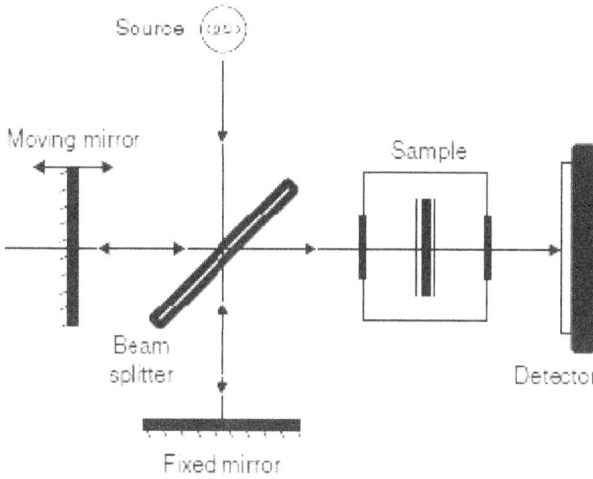

Figure 3.2. Schematic diagram of a Michelson interferometer. [Reprinted with permission from A. Subramanian and L. Rodriguez-Saona, *Infrared Spectroscopy for Food Quality Analysis and Control,* pp. 145–178, Elsevier, Amsterdam (2009). Copyright © 2009 Elsevier Inc. All rights reserved.]

two reflected beams from the mirrors recombine at the beam splitter. The difference in distance traveled by the two light beams is called the optical path difference (OPD) or optical retardation. The recombined beam passes through the sample and is finally detected by the detector.

Figure 3.3 shows the optical layout of a typical FTIR spectrometer [25]. In the FTIR spectrometer, the important component is the moving mirror. Accurate measurement of spectra is obtained by controlling the mirror position. The IR source emits infrared radiation. IR sources work based on the generation of heat due to resistance of the source to conduction of current. The resistance heats up the source to above 800°C, causing it to emit IR radiation. Because of the high operating temperatures, a cooling system is needed. Beam splitters serve to split and recombine the IR light waves in the interferometer. The function of the detector is to transduce the light intensity received by it to an electrical signal.

The detector response is measured at every zero crossing of the laser signal. It is also used for checking and aligning the optics in the interferometer. The spectrum of the sample obtained by Fourier transforming the interferogram is called the single-beam spectrum and represents the signal from the sample as well as that from the instrument and environment. The single-beam spectrum is against the background spectrum obtained without the sample, to obtain the actual spectrum of the sample.

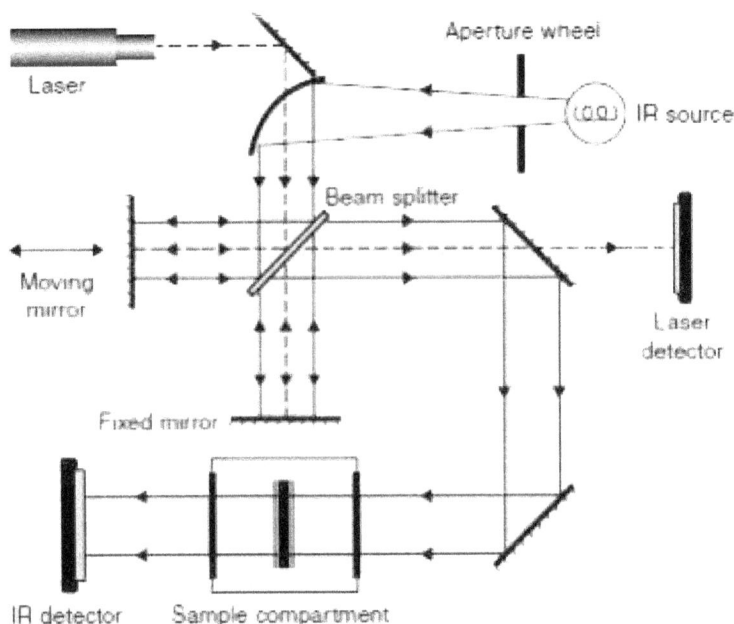

Figure 3.3. Optical layout of a typical FTIR spectrometer [Reprinted with permission from A. Subramanian and L. Rodriguez-Saona, *Infrared Spectroscopy for Food Quality Analysis and Control,* pp. 145–178, Elsevier, Amsterdam (2009). Copyright © 2009 Elsevier Inc. All rights reserved.]

3.1.3.   Attenuated Total Reflection (ATR)-FTIR

Attenuated total reflection (ATR) is added to FTIR to enhance the sampling techniques. The system is commonly known as ATR-FTIR [26,27]. This instrument helps to measure the sample cautiously with quantitative measurement [28–35]. ATR is more useful for strongly absorbing groups on thick samples, because the penetration depth of the momentary wave is only a few micrometers.

With internal reflection, infrared radiation is brought into contact with the sample with more refraction than the sample. The higher-refractive-index material is defined as the internal reflection element (IRE). The radiation is totally and internally reflected while incident radiation from the source is directed onto the interface between the sample and the IRE at an incident angle. In this process the momentary wave projects over a few micrometers beyond the surface of the IRE. The momentary wave is satisfied by the sample at the characteristic wavelength to an IR absorption band. This produces an attenuated total reflection (ATR) spectrum.

Even an opaque sample can be measured for spectra in ATR-FTIR. Figure 3.4 shows the infrared radiation path and contact type for ATR schematically [41]. Today's improved instruments help resolve the challenges of samples with greater complexity as well as expanding numbers of samples. However, the basic method of spectroscopic analysis has not changed substantially. Figure 3.4 represents the reflection of IR radiation during sample analysis.

ATR-FTIR is based on direct measurements, without sample preparation. However, the intensity of the infrared radiation is dependent on the sample depth, so it is necessary to correct the ATR-FTIR spectra with tools that are provided.

To increase the sensitivity, different forms of ATR sensing crystals have been studied for their performance in the detection of organic compounds in aqueous solutions [36–40].

The relation between ATR and transmission path and control type is not straightforward, as a correction is needed. In transmission, the optical path

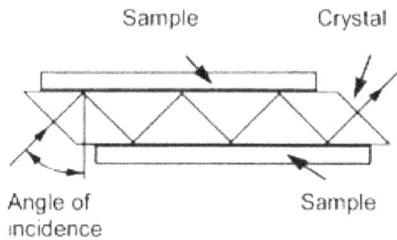

Attenuated Total Reflection

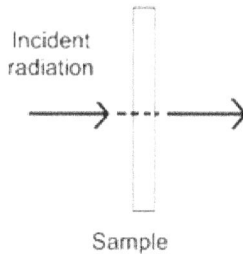

Transmission

Figure 3.4.   Schematic representation of the infrared radiation path and contact type for the different techniques. [Reprinted with permission from J. V. Gulminea, P. R. Janissek, H. M. Heisec, and L. Akcelrudd, *Polym. Testing* 21, 557 (2002). Copyright © 2002 Elsevier Science Ltd. All rights reserved.]

is the material thickness, whereas in ATR the depth of penetration (which is the counterpart of the optical path) is directly proportional to the wavelength.

The main advantage of ATR is the possibility of obtaining spectra directly from a polymer sheet, without any further sample preparation. Essentially, fundamental vibrations are detected with variations in intensity. ATR is a very sensitive technique and is especially suitable for surface analysis. The utmost care should be taken in sample manipulation, however, to avoid contamination.

In ATR the IR radiation is directed into the ATR crystal and is totally reflected within the crystal. The sample is brought into contact with the ATR crystal. The horizontal ATR design was invented by Messerschmidt [42]. It enables high throughput of the IR radiation and good reproducibility of the IR spectra without the need for optical realignment when changing crystals or samples. The reflection angle is simply determined by the grinding angle of the reflecting planes at the end of the crystal. These are coated with a reflecting metal layer. The sample is brought into contact with the crystal only at its upper side.

In most cases, the effective penetration depth is between 0.5 and 5 μm. This penetration depth is suitable for recording the IR spectra of organic materials and to realize maximum sensitivity for quantitative determination of concentrations.

The introduction of ATR objectives has reduced the need for elaborate sample preparation and dramatically decreased analysis time. A spectrum is obtained using a FTIR spectrometer fitted with an ATR-mode cell. The equipment is positioned in a laboratory maintained at $25 \pm 1°C$. A zinc selenide (ZnSe) ATR crystal, with suitable dimensions, angle of incidence, and refractive index is fitted to the spectrometer. The instrument is operated with resolution and the IR absorbency is analyzed between 700 and 4000 cm^{-1} for changes in the intensity of the sample peaks [43].

3.1.4.   FTIR and Analysis

Application of FTIR spectroscopy to polymer analysis consists of three parts: (1) sample preparation, (2) spectroscopic measurement, and (3) detection and identification.

For many analyses, generic fingerprinting, or determination of the chemical composition of a polymer, it is usually a simple matter to prepare a suitable thin-film specimen by a compression or microtome technique. To obtain good results, the sample should be as optically ideal as possible without devoting excessive time and effort to the preparation.

An FTIR spectrum is obtained from the spectrophotometer at room temperature using the potassium bromide (KBr) disc method for characterizing the polymer. To prepare the disc, the sample is dried and ground with KBr powder until they are in a well-mixed, powdered form. The powder is then pressed at 8 tons for 1 min to produce the disc. The sample is scanned at a wavenumber range of 4000 to 400 cm^{-1}.

Powdered samples mixed with solid KBr can be made up as a pellet. For sample concentration in the pellet of <0.1 µg, a microscope is used, Sample concentrations in the pellet of 0.1–0.2-µg require a microilluminator. A maximum of 8 µg of sample can be measured with a microcuvette, and samples up to 15 µg can be measured with an ATR crystal.

3.1.5.   Advantages

- FTIR is simple, fast, and precise [44].
- It requires a very small amount of sample for analysis.
- It is a valuable tool for in-situ monitoring of chemical reactions, because functional groups which are reacting or are produced during reaction show characteristic fingerprints in the IR spectrum.
- IR spectroscopy is useful for determining the functional groups present in a compound.
- It is most useful for obtaining qualitative information regarding the average oxidation rate.
- Within a class of compounds, infrared radiation absorbs essentially similar frequencies and intensities called group frequencies.

3.1.6.   Disadvantages

- FTIR frequently suffers from problems associated with sample preparation. Reflectance methods require suitable samples, whereas transmission methods demand samples with sufficient transparency to allow the penetration of IR beams.

3.2.   Raman Spectroscopy

Until about 1960, only infrared spectra were used in the characterization and analysis of polymers. Since then, Raman spectroscopy has increasingly been used to elucidate the structures of polymers [45]. Today, Raman spectroscopy remains in second place, behind FTIR, in the industrial laboratory.

Raman spectroscopy is a single-beam technique. Fluorescence is a time-dependent interference that appears as a broad band superposed to the Raman signal. It has also a strong influence over the whole intensity of the spectrum. Therefore, calibration procedures with external and internal standards are always needed to perform quantitative experiments. In Raman spectroscopy, the band intensity is the area of the band considered. These band areas are more significant than the band heights. The areas use more than one data point in the spectrum and are independent of spectral resolution [46]. However, the band measurement areas can be very complicated, because a series of overlapped narrow bands (Raman scattering) is superimposed on a very broad band (the fluorescence contribution). The interpretation of spectral data is often impeded by these effects. Numerical techniques such as deconvolution and curve fitting are widely used to extract information from such data.

Raman spectroscopy can provide quantitative information about structural details of polymers such as the chemical nature, stereoregularity, and state of order and orientation [47]. Some examples of quantitative applications of this technique include determination of crystallinity [48,49], conformational distribution in the polymer [50], and determination of composition in polymer isomers [51], blends of polymers, and copolymers [52–56].

The quantitative analysis of composition using Raman spectroscopy involves measurements of band intensities. The instrumental difficulty results in the measurement of absolute intensities and the overlapping of the bands in the Raman spectrum. However, in the Raman spectrum, the frequencies and relative intensities of the bands are functions of the sample. Hence the overall intensity of the whole recorded spectrum is determined by many instrumental factors [57] such as the intensity of the Raman source, optical alignment, and wavelength response.

3.2.1.   Principle and Operation

The Raman method relies on the detection of inelastically scattered photons, whereas most other spectroscopic methods, such as the near-infrared absorption used for remote sampling, rely on the detection of transmitted light through a sample stream. The nondestructive nature of Raman spectroscopy and the flexibility in sampling in principle provides an ideal method for analysis.

One of the major advantages of Raman spectroscopy is the ease of cell design, which arises from the fact that glass has a very weak Raman spectrum. This makes it an ideal material for cell construction or as windows in metal cells.

The study of polymers by Raman spectroscopy has for a long time been a desirable, but highly complex technique. Problems associated with fluorescence, exaggerated by the presence of additives in the polymer, has meant that the vibrational spectra of polymers have been mainly limited to infrared spectroscopy. Conventional Raman spectroscopy has been used only for materials that have been specially selected to be of low fluorescence and that have undergone extensive preparation before examination [59,60].

This is particularly unfortunate because the technique, with its high sensitivity to nonpolar species (such as C=C and C=S) which make up polymer chains and products, has obvious potential advantages over the complementary technique of IR absorption spectroscopy.

Recent advances, which combine the use of a near-infrared excitation source with Fourier-transform collection techniques [61–63] to produce Raman spectra, have enabled these problems to be substantially reduced, or overcome completely. This has triggered a reevaluation of the role of Raman spectroscopy in the study of polymers.

3.2.2.   Instrumentation

An FT-Raman spectrometer is similar to a conventional grating instrument, and the collection optics serves the same purpose. Figure 3.5 shows a diagram of a FT-Raman spectrometer [58]. In the instrument, the scattered light must be collected and then passed through a spectrometer. However, the half-angle divergence of the collected beam must neither exceed the resolution requirements of the interferometer nor pass through the limiting aperture stop of the interferometer. This condition is easily met for low-resolution experiments (1–4 cm^{-1}).

The main advantage of using FT-Raman spectroscopy lies in the negligible fluorescence which occurs when sampling from industrial reaction mixtures. These often contain low levels of impurities which have previously posed fluorescence problems when using visible laser excitation.

A mirror system gives the best spectra for small solid samples and powders; large samples occlude the Raman scatter. Their sensitivity is such that the Raman spectra of a wide range of polymers, in the as-received form, may be obtained in a few minutes, and without prior time-consuming optimization of sample placement in the spectrometer.

Raman scatter is processed by an interferometer to produce a FT-Raman spectrum. The positive features of FT-Raman spectroscopy include the need for little sample preparation and the rapid analysis time for most materials.

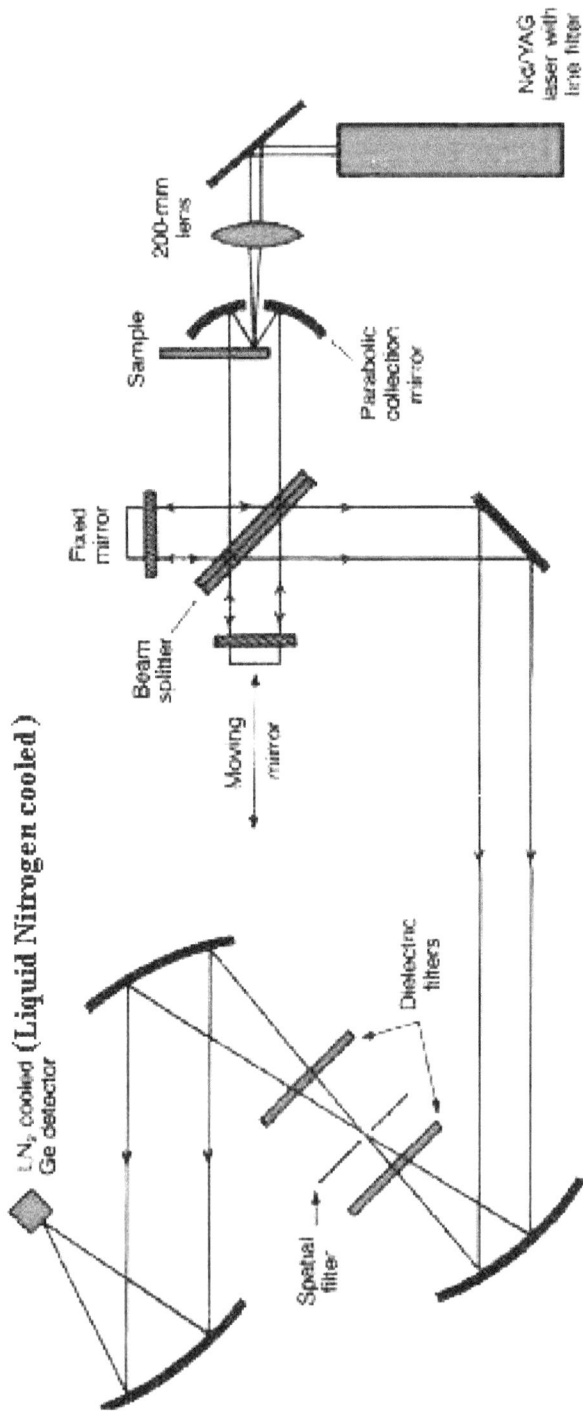

Figure 3.5. Diagram of a FT-Raman spectrometer. [Reprinted with permission from B. Chase, *Anal. Chem.* 59, 14 (1987). Copyright © 1987 American Chemical Society.]

Raman spectroscopy is complementary in nature to infrared spectroscopy [64], and can provide information from functional groups with vibrational modes that are weak or unresolvable by FTIR.

3.2.3.   Advantages

- Raman spectroscopy is especially suited to the study of polymers because of its sensitivity to the structure of molecules containing non-polar species [C–O–C] which commonly make up the backbone of synthetic polymers.
- No special sample preparation is required. (Sample preparation might change the conformational content or crystallinity.)
- It is a scattering technique. It does not suffer from the sampling limitations. The technique can be used for routine analysis.

3.2.4.   Disadvantages

- The occurrence of sample fluorescence may create problems [65].
- The Raman scattered signal is relatively weak.
- Samples may photodegrade.

3.3.   Nuclear Magnetic Resonance Spectroscopy (NMR)

The study of solid polymers has developed significantly with the application of nuclear magnetic resonance spectroscopy in recent years. Analytical techniques can rarely compete with the versatility of NMR. It is an essential tool for the structural elucidation of the molecular components of polymers. Structural investigations may involve the identification of polymers, either in blends or after purification. NMR has proved to be of great importance in many aspects, and its use has expanded rapidly [66–71].

3.3.1.   Principle

NMR is based on nuclei which possess both magnetic and angular moments. Nuclei which have odd mass number or odd atomic number interact with an applied magnetic field M_0. The applied magnetic field yields $2I + 1$ (where I is the nuclear spin quantum number) energy levels with separation of

$$\Delta E = h\omega = \gamma h M_0 \qquad (3.2)$$

where h is the planck constant divided by 2π, ω is the Larmor frequency of nuclear precession, and γ is the gyromagnetic ratio. The interaction of a single spin with the magnetic field (in the range 10^6–10^8 Hz) is described by a Zeeman Hamiltonian H_z:

$$H_z = -\gamma h M_0 I_z \qquad (3.3)$$

where I_z is the z component of the spin angular momentum operator I (in the direction of the applied field). Spectroscopic detection of these energy levels is possible when transitions between them are induced by an alternating magnetic field $M_1(\omega t)$ of frequency ω (perpendicular to the static field M_0) which satisfies the resonance condition $\omega = \gamma M_0$.

The properties of multispin systems are determined by different types of interactions. They can be described by a Hamiltonian H:

$$H = H_z + H_D + H_Q + H_J + H_\delta \qquad (3.4)$$

where the Hamilitonian terms are described as follows:

H_z = Zeeman intereaction with the applied field
H_D = direct dipole–dipole interaction with other nuclei
H_Q = quadrupolar interaction (for nuclei with $I > \frac{1}{2}$)
H_J = chemical-shift interaction
H_δ = indirect (electron-coupled) spin–spin couplings to other nuclei

Contributions of the four last terms in Eq. (3.4) depend on the physical state. In the solid state, the strong dipolar and quadrupolar terms are usually dominant, and the weak interactions such as chemical shift and spin–spin coupling are obscured.

In contrast, the dipolar and quadrapolar interactions in liquids average to zero, giving rise to high-resolution spectra in which chemical shifts and J couplings can be observed. Furthermore, rapid motions in solutions average the above-mentioned tensors to scalar quantities. According to these observations, NMR studies can be classified into two major domains: broad-line (low-resolution) studies of the solid state and sharp-line (high-resolution) studies in the liquid state.

However, it is found that a combination of techniques, such as proton dipolar decoupling (removes the dipolar interactions), magic-angle spinning (reduces the chemical shift tensor to the isotropic chemical-shift value), and cross-polarization (increases the sensitivity of rare spins, such as ^{13}C), applied to a solid-state material, results in sharp lines for ^{13}C nuclei in the solid state [72]. Thus, the observation of narrow lines or high-resolution NMR in the solid state is possible.

The NMR phenomenon may be explained by both classical and quantum mechanical descriptions, and each has certain advantages. Atomic nuclei behave as spinning bodies due to the presence of charge and mass, and for NMR should have magnetic moments and angular momentum. The NMR phenomenon is not exhibited with nuclei that have no nuclear spin even though they have mass and atomic number. These include the common isotopes ^{12}C, ^{28}Si, and ^{16}O.

Nuclei with odd mass numbers and even nuclear charge numbers have spin quantum numbers I which are odd integral multiples of $\frac{1}{2}$, whereas those with even mass numbers and odd nuclear charge have integral values of I. With the help of the magnetic moment (μ) and angular momentum by a factor (γ) called the magnetogyric ratio, the equation becomes

$$\mu = \gamma hI/2\pi \qquad\qquad (3.5)$$

where h is Planck's constant.

NMR signals for different nuclei are dependent on I, γ, and the natural abundance of the isotope. Nuclei with $I = \frac{1}{2}$ produce spectra with the narrowest lines. Nuclei with $I > \frac{1}{2}$ possess electric quadrupole moments which can significantly shorten relaxation times and broaden resonance lines. For this reason these nuclei are generally less suitable for traditional NMR experiments. The magnetogyric ratio γ is a fundamental constant of the nucleus, which determines the frequency and inherent sensitivity of the nucleus to the NMR experiment. In general, spin $\frac{1}{2}$ nuclei with a high γ value and reasonable natural abundance, such as ^{1}H, ^{19}F, and ^{31}P, are easily observed by NMR spectroscopy.

Nuclei with $I = \frac{1}{2}$ have two spin states of magnetic moment aligned either with or against the applied magnetic field ($m = +\frac{1}{2}$ or $m = -\frac{1}{2}$). This forms the basis of NMR experiments. These spin states are degenerate in the absence of a magnetic field. However, they correspond to different states of potential energy in the presence of a uniform magnetic field (M_0).

As given by the Boltzmann population distribution, the separation of the energy levels is small, and at equilibrium there is only a slight excess of nuclei in the lower-energy spin state. NMR experiments are dependent on small populations to produce signal. Compared to other spectroscopic techniques, NMR is relatively insensitive. Once the energy transfer occurs into the spin system, the process by which the nuclei reestablish a Boltzmann population distribution is called spin–lattice relaxation.

3.3.2.   Solvents

Solvents for the NMR study of polymers are still something of an art. To choose solvents, certain criteria have to be ensured:

1. Absence of interfering lines. This requirement is no different from that of NMR generally and calls for no special comment.
2. Possibly low or moderate viscosity is necessarily important even though the resulting polymer solution may be highly viscous.
3. Moderately high boiling point. Low-viscosity solvents such as CS_2 and $CHCl_3$ (or $CDCl_3$) are not suitable above 100°C because of bubbling and refluxing.
4. Absence of paramagnetic impurities. These may broaden lines through shortening nuclear relaxation times.
5. Careful filtration. This is desirable to avoid cross-linked gels and remove impurities.

Polymer spectra are a function of local segmental and side-chain motion. They are nearly independent of molecular weight and therefore of the macroscopic viscosity of the solution. Line widths are also nearly independent of polymer concentration in lower concentration ranges.

3.3.3.   Instrumentation

In modern NMR spectrometers, as shown in Figure 3.6, the magnetic field is generated by a superconducting magnet. Samples are placed in a strong, homogenous magnetic field and radio-frequency (RF) electromagnetic energy is applied. The spectrometer console provides at least three RF channels which are transmitted to the probe. The observed RF for the nucleus of interest is supplied by the observe channel. It also receives the signal back from the sample. Nuclei absorb particularly that energy at sharply defined frequencies. The signal is detected, amplified, and presented as a frequency spectrum. The resonance frequency depends on its local electronic environment and is measured against a reference peak in the sample. The area under

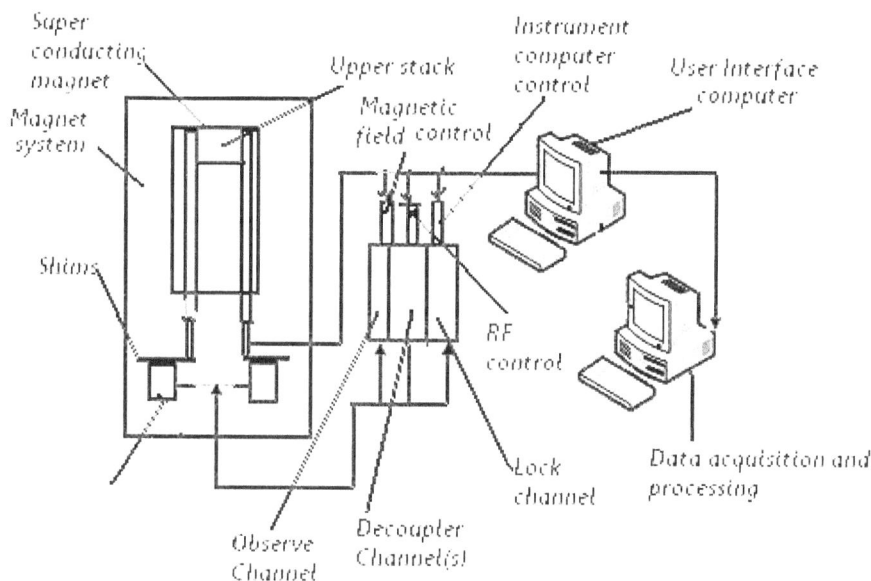

Figure 3.6. Schematic of a typical NMR spectrometer.

the resonance line can be integrated to give quantitative information about the relative number of nuclei giving rise to each peak.

The splitting pattern indicates structural information about the neighboring nuclei. This ensures long-term stability of high-resolution NMR experiments. Modern instrumental methods employ two computers, using one for data acquisition and processing and the other for spectrometer control.

3.3.4.   Chemical Shifts

NMR spectroscopy is an analytical technique used extensively in molecular structure determination of polymeric material. It depends on the magnetic properties of atomic nuclei. Each nucleus is surrounded by an electron cloud. The motion of electrons creates a secondary magnetic field which tends to oppose the applied magnetic field. Nuclei of a given type do not all exhibit transitions at exactly the same frequency. The chemical-shift differences arise because the electron density varies depending on the electronegetivity of the atoms attached to the nucleus and the single or multiple bonds in which it is involved.

The higher the electron density near the nucleus is, the greater will be the secondary magnetic field or shielding effect. The effect of the nearest neighboring atoms or functional groups is usually most important.

The chemical shift is the dependence of the resonance frequency of a given nucleus on its special chemical environment in the molecule. It establishes NMR as a fundamental tool for structure determination. Spectra may be obtained even with low sample concentration. Also, nuclei may be observed that have very low natural abundance. The resolution of structural features is greatly improved over other forms of analysis.

The majority of polymer studies have utilized proton measurements, because its nucleus is most amenable to NMR observation and the proton has the smallest range of chemical shifts. The other nucleus common to virtually all polymers is ^{13}C.

Earlier, NMR studies of solid polymers provided weld-line-type spectra [73–75]. The most prominent characteristic of polymer spectra, particularly structurally irregular polymers rather than those composed of exactly repeating units, is that the lines are broadened compared to those of small molecules of analogous structure [76–78]. The reasons have been clearly understood only recently, but it appears that the line broadening arises from three causes: (1) dipolar broadening, (2) multiple transitions, and (3) multiple chemical shifts.

Many interactions give rise to a multiplicity of lines, which cannot be resolved but whose envelope is seen as a continuous curve. In most solids, the NMR signals result in a single broad line, and pairing of protons does not occur to any significant degree. In most rigid solids, however, the distribution of local field strength is the half-height width of the resonance.

In NMR, both solid state and solution, measurement of the spin–lattice and spin–spin nuclear relaxation is included. This provides further insight into the motion and interactions of polymer chains [79–90]. In polymer spectra, to achieve spectral simplification and signal-to-noise improvement, principal use has been made of deuterium substitution, double resonance, computer simulation, and spectrum accumulation using a computer to average transients.

By its very nature, NMR is sensitive on the scale of molecular moieties. In solids, NMR yields information on composition, molecular mobility, and molecular order [91,92]. In addition, diffusion of proton magnetization through the network of homonuclear dipolar couplings can be exploited to investigate structures in the range of tens of nanometers [93–100].

3.3.5. Advantages

- NMR spectroscopy is an indispensable tool in polymer testing. Its applications include the structure determination of polymer and supramolecular systems that are not crystalline in the traditional sense [101].

- NMR spectroscopy can be carried out at very high magnetic fields. With ^{1}H NMR, it is possible now to use frequencies up to 2.4 GHz [102].
- NMR spectra can confirm the structure and conformation of polymer solutions.
- It is a powerful method for large molecules, including polymers.
- It can provide rapid preliminary analytical examination of small quantities of polymers that are sufficiently soluble in the usual solvents, for example, deuterochloroform and hexadeuterodimethylsulfoxide.
- It provides information concerning the symmetry of uncertain structures or stereochemical configurations.

3.3.6. Disadvantages

- In solid-state spectra, the general feature of providing information related to structure and conformation cannot be resolved.
- Despite the success of NMR spectroscopy, which results from its extreme site selectivity, the inherently low signal intensity remains a serious drawback which limits its application.

References

1. H. V. Boeing, *Polyolefins: Structure and Properties,* pp. 93 and 132, Elsevier, Amsterdam (1966).
2. C. Jén-Yüan, *Determination of Molecular Weights of High Polymers,* p. 27, Israel Program for Scientific Translations, Jerusalem (1963).
3. F. Bueche, *Physical Properties of Polymers,* p. 222, Interscience, New York (1962).
4. J. M. G. Cowie, *Polymers—Chemistry and Physics of Modern Materials,* p. 263, Intertext Books, London (1973).
5. S. C. Hodgson, S. W. Bigger, and N. C. Billingham, *J. Chem. Educ.* 78, 555 (2001).
6. J. M.,Chalmers, L. Croot, J. G. Eaves, N Everall, W. F. Gaskin, J Lumsdon, and N. Moore, *Spectrosc. Int. J.* 8, 13 (1990).
7. J. M. Chalmers and N. J. Everall, *Macromol. Symp.* 94, 33 (1995).
8. F. S. Parker, *Application of Infrared Raman and Resonance Raman Spectroscopy in Biochemistry,* pp. 354–356, Plenum Publishing, New York (1983).
9. P. T. T. Wong, S. M. Golstein, R. C. Grekin, T. A. Godwin, C. Pivik, and B. Rigas, *Cancer Res.* 53, 762 (1993).
10. P. R. Griffiths and J. A. de Haseth, *Fourier Transform Infrared Spectroscopy,* John Wiley, New York (1986).

11. P. R. Griffiths, Fourier transform infrared spectroscopy, *Science* 222, 297 (1983).

12. J. E. Katon, *Vib. Spectrosc.* 7, 201 (1994).

13. N. B. Colthup, L. H. Daly, and S. E. Wiberley, *Introduction to Infrared and Raman Spectroscopy,* Academic Press, New York (1964).

14. P. R. Griffiths, *Chemical Infrared Fourier Transform Spectroscopy,* John Wiley, New York (1975).

15. L. D. Esposito and J. L. Koenig, in J. Ferraro and L. J. Basile (eds.), *Fourier Transform Infrared Spectroscopy,* Vol. 1, Academic Press, New York (1978).

16. H. Ishida (ed.), *Fourier Transform Infrared Characterization of Polymers,* Plenum Press, New York (1987).

17. I. J. Spiro and M. Schlessinger, *Infrared Technology Fundamentals,* Marcel Dekker, New York (1989).

18. A. Garton, *Infrared Spectroscopy of Polymer Blends, Composites and Surfaces,* Hanser, New York (1992).

19. R. W. Sebesta and G. G. Johnson, Jr., *Anal. Chem.* 44, 260 (1972).

20. L. H. Fan, C. P. Hu, Z. P. Zhang, and S. K. Ying, *J. Am. Chem. Soc.* 59, 1417 (1996).

21. S. Keskin and S. Ozkar, *J. Appl. Polym. Sci.* 81, 918 (2001).

22. D. Kincal and S. Ozkar, *J. Appl. Polym. Sci.* 66, 1979 (1997).

23. B. Stwart, in D. J. Ando (ed.), *Biological Applications of Infrared Spectroscopy,* pp. 25–30, John Wiley, Greenwich, U.K. (1997).

24. P. T. T. Wong and H. H. Mantsch, *Biomol. Spectrosc.* 1057, 49 (1989).

25. A. Subramanian and L. Rodriguez-Saona, *Infrared Spectroscopy for Food Quality Analysis and Control,* pp. 145–178, Elsevier, Amsterdam (2009).

26. J. Fahrenfort, in A. Mangini (ed.), *Proceedings IV International Meeting on Molecular Spectroscopy,* Vol. 2, p. 437, Pergamon Press, London (1962).

27. J. Fahrenfort, in A. Mangini (ed.), *Proceedings IV International Meeting on Molecular Spectroscopy,* Vol. 2, p. 437, Pergamon Press, London (1962).

28. A. L. Smith, *Applied Infrared Spectroscopy: Fundamentals, Techniques and Analytical Problem-Solving,* Wiley Interscience, New York (1986).

29. D. Rivera, P. E. Poston, R. H. Uibel, and J. M. Harris, *Anal. Chem.* 72, 1543 (2000).

30. D. Rivera and J. M. Harris, *Anal. Chem.* 73, 411 (2001).

31. F. G. Haibach, A. Sanchez, J. A. Floro, and T. M. Niemczyk, *Appl. Spectrosc.* 56, 398 (2002).

32. L. Han, T. M. Niemczyk, D. M. Haaland, and G. P. Lopez, *Appl. Spectrosc.* 53, 381 (1999).

33. R. Howley, B. D. MacCraith, K. O'Dwyer, H. Masterson, P. Kirwan, and P. McLoughlin, *Appl. Spectrosc.* 57, 400 (2003).

34. R. Howley, B. D. MacCraith, K. O'Dwyer, P. Kirwan, and P. McLoughlin, *Vib. Spectrosc.* 31, 271 (2003).

35. J. V. Gulmine, P. R. Janissek, H. M. Heise, and L. Akcelrud, *Polymer Testing* 21, 557 (2002).

36. A. Messica, A. Greestein, and A. Katzir, *Appl. Opt.* 35, 2274 (1996).

37. S. Simhony, A. Katzir, and E. W. Kosower, *Anal. Chem.* 60, 1908 (1988).
38. R. Gobel, R. Krska, R. Kellner, J. Kastner, A. Lambercht, M. Tacke, and A. Katzir, *Appl. Spectrosc.* 49, 1174 (1995).
39. P. H. Paul and G. Kychakoff, *Appl. Phys. Lett.* 51, 12(1987).
40. K. Newby, W. M. Reichert, J. D. Andrade, and R. E. Benner, *Appl. Opt.* 23, 1812 (1984).
41. J. V. Gulminea, P. R. Janissek, H. M. Heisec, and L. Akcelrudd, *Polym. Testing* 21, 557 (2002).
42. R. G. Messerschmidt, U.S. Patent 4,730,882.
43. A. C. Ruddy, G. M. McNally, G. Nersisyan, W. G. Graham, and W. R. Murphy, *J. Plastic Film Sheeting* 22, 103 (2006).
44. B. Foster, *Am. Lab.* 11, 42 (2001).
45. H. W. Siesler and K. Holland-Moritz, *Infrared and Raman Spectroscopy of Polymers,* Marcel Dekker, New York (1980).
46. P. Hendra, C. Jones, and G. Warnes, *Fourier Transform Raman Spectroscopy, Instrumental and Chemical Applications,* Ellis Horwood, Chichester, U.K. (1991).
47. A. Fawcett (ed.), *Polymer Spectroscopy,* John Wiley, Chichester, U.K. (1996).
48. D. L. Gerrard and W. F. Maddams, *Appl. Spectrosc. Rev.* 22, 251 (1986).
49. G. R. Strobl and W. Hagedorn, *J. Polym. Sci. Polym. Phys. Ed.* 16, 1181 (1978).
50. J. C. Rodriguez-Cabello, L. Quintanilla, and J. M. Pastor, *J. Raman Spectrosc.* 25, 335 (1994).
51. K. D. O. Jackson, M. J. R. Loadman, C. H. Jones, and G. Ellis, *Spectrochim. Acta* 46A, 217 (1990).
52. J. K. Agbenyega, G. Ellis, P. J. Hendra, W. F. Maddams, C. Passingham, H. A. Willis, and J. Chalmers, *Spectrochim. Acta* 46A, 197 (1990).
53. K. P. J. Williams and S. M. Manson, *Spectrochim. Acta* 46A, 197 (1990).
54. J. P. Tomba, J. M. Carella, J. M. Pastor, and M. R. Fernández, *Macromol. Rapid Commun.* 19, 413 (1998).
55. S. Hajatdoost, M. Olsthoorn, and J. Yarwood, *Appl. Spectrosc.* 51, 1784 (1997).
56. S. Hajatdoost, and J. Yarwood, *Appl. Spectrosc.* 50, 558 (1996).
57. P. Hendra, C. Jones, and G. Warnes, *Fourier Transform Raman Spectroscopy, Instrumental and Chemical Applications,* Ellis Horwood, Chichester, U.K. (1991).
58. B. Chase, *Anal. Chem.* 59, 14 (1987).
59. S. W. Cornell and J. L. Koenig, *Macromolecules* 2, 546 (1969).
60. K. Kurosaki, *Int. Polym. Sci. Technol.* 15, 601 (1988).
61. B. Chase, *Anal. Chem.* 59, 881A (1987).
62. V. M. Hallmark, C. G. Zimba, J. D. Swalen, and J. F. Rabolt, *Spectroscopy* 2, 40 (1987).
63. P. J. Hendra and H. Mould, *Int. Lab.* 18, 34 (1988).
64. D. I. Bower and W. F. Maddams, *The Vibrational Spectroscopy of Polymers,* Cambridge University Press, Cambridge, U.K. (1989).
65. H. Baranska, A. Labudzinska, and J. Terpinski, *Laser Raman Spectroscopy— Analytical Applications,* Ellis Horwood, Chichester, U.K. (1987).

66. F. A. Bovey and G. V. D. Tiers, *Fortschr. Hochpolym.* 3, 139 (1963).
67. D. W. McCall and W. P. Slichter, in B. Ke (ed.), *Newer Methods of Polymer Characterization,* Wiley-Interscience, New York (1964).
68. F. A. Bovey, F.A., Nuclear magnetic resonance, in *Encyclopedia of Polymer Science and Technology,* Vol. 16, Wiley-Interscience, New York (1968).
69. K. C. Ramay and W. S. Brey, Jr., *J. Macromol. Sci. C* 1, 263 (1967).
70. H. A. Willis and M. E. A. Cudby, *Appl. Spectrosc. Rev.* 1, 237 (1968).
71. J. C. Woodbrey, in A. D. Ketley (ed.), *The Stereochemistry of Macromolecules,* Vol. 3, Marcel Dekker, New York (1968).
72. J. Schaefer and E. O. Stejskal, *J. Am. Chem. Soc.* 98, 1031 (1976).
73. W. P. Slichter, *Fortschr. Hochpolym. Forsch.* 1, 35 (1958).
74. J. G. Powles, *Polymer* 1, 219 (1960).
75. J. A. Sauer and A. E. Woodward, *Rev. Mod. Phys.* 32, 88 (1960).
76. M. Saunders and A. Wishnia, *Ann. N. Y. Acad. Sci.* 70, 870 (1958).
77. A. Odajima, *J. Phys. Soc. Japan* 14, 777 (1959).
78. F. A. Bovey, G. V. D. Tiers, and G. Filipovich, *J. Polym. Sci.* 38, 73 (1959).
79. A. W. Nolle and J. J. Billings, *J. Chem. Phys.* 30, 84 (1959).
80. J. G. Powles and K. Luszczynski, *Physica* 25, 455 (1959).
81. G. A. Powles, A. Hartland, and A. E. Kail, *J. Polym. Sci.* 55, 361 (1961).
82. E. G. Kontos and W. P. Slichter, *J. Polym. Sci.* 61, 61 (1962).
83. W. P. Slichter and D. D. Davis, *J. Appl. Phys.* 34, 98 (1963).
84. W. P. Slichter and D. D. Davis, *J. Appl. Phys.* 35, 3103 (1964).
85. D. W. McCall and E. W. Anderson, *Polymer* 4, 93 (1963).
86. J. G. Powles, J. H. Strange, and D. J. Sandiford, *Polymer* 4, 401 (1963).
87. J. G. Powles, B. I. Hunt, and D. J. Sandiford, *Polymer* 5, 585 (1964).
88. W. P. Slichter, *J. Polym. Sci. C* 14, 33 (1966).
89. D. W. McCall, D. C. Douglass, and D. R. Falcone, *J. Phys. Chem.* 71, 998 (1967).
90. W. P. Slichter and D. D. Davis, *Macromolecules* 1, 47 (1968).
91. H. W. Spiess, *Advances in Polymer Science, Vol. 66,* Springer-Verlag, Berlin (1985).
92. M. Mehring, *Principles of High Resolution NMR in Solids,* Springer-Verlag, Berlin (1983).
93. N. Bloembergen, *Physica* 15, 386 (1949).
94. M. Goldman and L. Shen, *Phys. Rev.* 144, 321 (1961).
95. T. T. P. Cheung and B. C. Gerstein, *J. Appl. Phys.* 52, 5517 (1981).
96. P. Caravatti, P. Neuenschwander, and R. R. Ernst, *Macromolecules* 18, 119 (1985).
97. P. Caravatti, M. H. Levitt, and R. R. Ernst, *J. Magn. Reson.* 68, 323 (1986).
98. P. Caravatti, P. Neuenschwander, and R. R. Ernst, *Macromolecules* 19, 1889 (1986).
99. K. J. Packer, J. M. Pope, R. R. Yeung, and M. E. A. Cudby, *J. Polym. Sci., Polym. Phys. Ed.* 22, 589 (1984).
100. J. R. Havens and D. L. VanderHart, *Macromolecules* 18, 1663 (1985).
101. H. W. Spiess, *J. Polym. Sci. A* 42, 5031 (2004).
102. M. B. Kozlov, J. Haase, C. Baumann, and A. G. Webb, *Solid State Nucl. Magn. Reson.* 28, 64 (2005).

Chapter 4

Thermal Analysis and Degradation

Devices for the thermal analysis of polymers were described as early as the 1950s [1], and today they find wide use in polymer science. Polymers encounter elevated temperatures at almost every step in the manufacturing, compounding, and processing stages, and even in service, so a great deal of information is needed about their thermal behavior.

Instruments to measure temperature by scanning with automatic control and high resolution first became available commercially in the 1960s [2]. Instrumental methods research led to inventions of a variety of thermal analysis devices with abilities beyond measurement of thermodynamic properties. Thermal analysis of nonthermodynamic properties extends the temperature-scanned variables from thermodynamic properties to other polymer properties.

Thermal analysis is considered an advance from qualitative to quantitative measurement. Thermal fractionation is an improved step in modern methods to understand polymer structure and its relation to performance. Instrumental methods are available for identification and characterization to determine many polymeric material characteristics. Thermal analysis can be carried out with the help of thermogravimetric analysis (TGA), differential scanning calorimetry (DSC), and differential thermal analysis (DTA). Polymer testing demonstrates the lengthy chain and high-molecular-weight nature of polymers.

DOI: 10.5643/9781606502440/ch4

4.1.   Thermogravimetric Analysis

Thermogravimetric analysis (TGA) is the art and science of weighing substance while they are being heated. It offers the possibility of studying the complete thermal decomposition behavior of a polymer substance. It is simple and gives a weight–temperature relation directly. TGA results can be surprisingly good, and many of the apparent limitations of thermogravimetric measurements can be effectively overcome.

TGA is one of the most commonly used thermal analyses techniques for the characterization of polymers. It provides quantitative results about the weight loss of a sample as a function of temperature or time. It gives basic information about the thermal properties of the material and its composition. Thermogravity provides details of thermal decomposition processes, which in turn facilitate estimation of sample structure and composition [3–6].

4.1.1.   Principle

Isothermal or static thermogravimetry is an old art. The new instruments facilitate studies by dynamic thermogravimetry in which the sample is heated at a uniform rate. Hence thermogravimetry has increased in recent years because of the commercial availability. Care must be taken to avoid errors, which may include the effect of changing air buoyancy and convection, the measurement of temperature, and the effects of atmosphere, heating rate, and heat of reaction. Thermogravimetric analysis is an invention of microbalance and minimization of effect of purge gas on the balance sensitivity during quantitative analysis.

TGA uses a sensitive balance. The thermobalance facilitates studies of a sample which is subjected to conditions of continuous increase in temperature, usually linear with time. The weight loss during the sample decomposition is measured with a thermogravimetric analyzer. It uses either inert or nitrogen gas or air atmosphere for a given specific chemical measured. A TGA experiment depends mainly on initial weight and temperature, heating rate, and gas flow (air, nitrogen, argon, etc.). The experimental conditions of every experiment are carried out with TGA to design, optimize, and test the sample. TGA experiments are performed using crucibles made of alumina (Al_2O_3) or platinum in the temperature range 23–930°C.

4.1.2.   Instrumentation

The weight loss of a polymer as a function of time or temperature is commonly determined by the technique of TGA. Weight loss of a polymer due to thermal degradation is an irreversible process. This thermal degradation is largely related to oxidation, whereby the molecular bonds of a polymer are attacked by oxygen molecules. Figure 4.1 shows a schematic diagram of a thermogravimetric analyzer [7].

The TGA equipment is able to measure the sample weight loss as a function of temperature or heating time with an accuracy of ±1 μg. The sample temperature is measured with an accuracy of ±0.1°C [8]. The thermogravimetric analyzer is coupled to a main console to record measurements. The analyzer contains a semi-micro balance designed to measure weight loss as a function of temperature or as a function of time at a given temperature. However, the main console is programmed to determine the mode of operation and the rate of heating at the analyzer.

The thermogravimetric analyzer is capable of heating to a temperature of 1200°C with any linear heating rate between 0.5 and 30°C/min. The analyzer operates on a null balancing principle. The balance beam is maintained in a reference position. The beam is balanced by a taut band electric movement. The overall sensitivity is about 10 μg.

Thermogravimetric analysis of a polymeric material provides a plot of weight change against temperature or time in a controlled dynamic or static

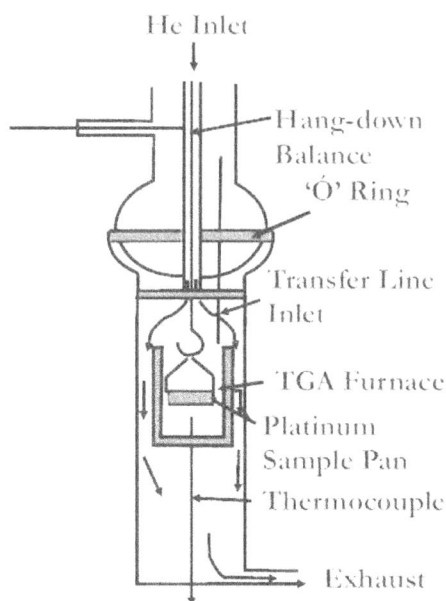

Figure 4.1. Schematic diagram of a TGA. [Reprinted with permission from A. P. Snyder, A. Tripathi, J. P. Dworzanski, W. M. Maswadeh, and C. H. Wick, *Anal. Chim. Acta* 536, 283 (2005). Copyright © 2005 Elsevier B.V. All rights reserved.]

temperature environment. It is used fairly widely as a characterization technique for polymeric materials [9–11].

Thermogravimetric analysis ultimately affords a record of residual weight fraction versus temperature for a sample heated at a fixed rate under a particular set of other experimental conditions [12]. Ideally, a single thermogram is equivalent to a very large family of comparable isothermal volatilization curves and, as such, constitutes a rich source of kinetic data for volatilization. While it is often quite difficult to realize this potential and even more difficult to apply the resulting laboratory test data to practical problems involving different conditions of geometry, function, and environment, the determination is worth undertaking as a first step.

4.1.3. Advantages

- TGA helps to identify a complete polymer decomposition profile by determining weight losses of different molecular fragments separating out by a heating a small specimen from room temperature at a controlled heating rate until it is fully combusted into ashes.

- It helps to develop new products with controlled molecular structure and tailor-made products with specific melting, crystallization, and/or decomposition characteristics.

- Polymer products based on the different crystallization behaviors of the various species present can be identified.

- The thermogravimetric technique has a number of advantages over the more conventional isothermal method of obtaining kinetic parameters, such as energy of activation and order of reaction.

4.1.4. Disadvantage

- It is a sample-destructive technique.

4.2. Differential Scanning Calorimetry

Today thermal analysis is usually rapid, sensitive, and computerized, and includes a wide choice of techniques over a broad range of temperature.

Nevertheless, the ultimate goal of thermal analysis for analytical purposes or in the thermal preparation of solids is still:

- To obtain thermoanalytical curves and kinetic parameters which will be independent of the sample mass (especially when the requirements of satisfactory sampling do not allow the sample size to be decreased at will), of the shape of the crucible, and of the location of the temperature sensor, so that they will be, above all, characteristic of the samples studied and not of the experimental arrangement.

- To have more reproducible thermal preparations, on both the laboratory and the industrial scales.

- To disentangle the various phenomena occurring in complex thermolyses (i.e., those leading to several gaseous or solid products) by simple and safe processing of the thermal analysis data, and to obtain "meaningful" and useful apparent energies of activation [13].

Differential scanning calorimetry (DSC) is fundamentally a new approach to quantitative thermal analysis using micro samples. Unlike conventional techniques, an isothermal calorimeter measuring the thermal energy difference required to keep sample and reference at a constant temperature is used. Conventional thermal analytical techniques measure the temperature (or some other property) difference between sample and reference developed upon heating (or cooling) at a fixed rate, and so they are not isothermal. DSC is potentially capable of satisfying the strict requirements of isothermal crystallization kinetic studies.

DSC is able to fractionate polymer products based on the different crystallization behaviors of the various species present during its cooling and isothermal cycles. It is used to characterize the thermal behavior of polymers [14]. It can also be used to characterize the thermal behavior of semicrystalline, modified polymers by indicating the presence of specific crystalline components [15].

DSC gives a measure of the difference in the rates of heat absorption by a sample with respect to an inert reference as the temperature is raised at a constant rate. This contrasts with conventional differential thermal analysis, in which the differential temperature caused by, but only indirectly related to, heat changes in the sample is monitored. Measurements of heats of reaction in a sample whose temperature is scanned should give useful information on the reaction kinetics if it is assumed that the heat of reaction is directly proportional to the extent of the reaction [16,17]. This is an assumption which is reasonable for simple reactions but not obviously valid for the complex cross-linking reactions which take place as an epoxy resin polymerizes.

4.2.1.   Principle

When heated to high enough temperatures, all materials undergo physical or chemical changes, whether they are bulk materials or thin films. These changes alter the enthalpy and/or heat capacity of the material, which in turn results in the release or the absorption of heat. By determining the instantaneous rate of heat flow (namely, the power), differential scanning calorimetry provides quantitative thermodynamic and kinetic information about the physical and chemical changes occurring in the material.

DSC yields a thermogram of rate of heat absorption, dH/dt, as a function of temperature, T. Because a constant heating rate, dT/dt, is used, this rate of heat absorption is proportional to the specific heat of the sample. In the absence of chemical reaction, therefore, a second-order transition is shown as a discontinuity in the thermogram. Second-order transitions associated with the softening point in cross-linked networks have been investigated by Gordon et al. [18] using a dynamic mechanical method, and the temperature of the transition is shown to increase with the extent of cure. If it approaches the cure temperature, further reaction must be controlled by diffusion processes in the glassy state.

DSC is followed by a more detailed treatment of the quantitative aspects of three applications that have widespread usage in the polymer industry. The technique provides the values of temperature (°C) and enthalpy (J/g) corresponding to the following transitions: (1) the glass transition; (2) heats of reaction (curing); and (3) crystallinity

DSC can be performed in either a heating mode, to obtain melting peaks associated with endothermic reactions, or in a cooling mode, to obtain crystallization peaks associated with exothermic reactions. Using the heating mode of testing, the first-order transitions, associated with melting of the crystalline fractions of the polymer, are accompanied by endothermic signals which are easily characterized by DSC analysis. The parameters recorded from the test are the enthalpy associated with the endothermic reaction and the peak temperature at which the crystalline components melt.

DSC is used for the measurement of polymer response to a programmed change of temperature. The parameter monitored may range from the familiar, such as a mass or a dimension, to some less common optical or acoustical property. DSC measures heat capacities (C_p).

Isothermal and dynamic (cooling) crystallizations of the sample are normally observed with DSC, since differential thermal analysis (DTA) is not suitable for kinetic analysis.

4.2.2.   Instrumentation

Differential scanning calorimetry uses a symmetrical double-staged cell design with quantitative heat flow measurements of the cell based on the heat flux or power consumption. In the DSC the sample and the reference are thermally insulated from one another, and each is provided with its own individual heater, as shown schematically in Figure 4.2. The instrument is based on the "null balance" principle of measurement, in which the energy absorbed or evolved by the sample is compensated by adding or subtracting an equivalent amount of electrical energy to the heater located in the sample holder. In practice, this is achieved by comparing the signal from a platinum resistance thermometer in the sample holder with that from an identical sensor in the reference holder. (These thermometers need to be calibrated, since the true sample temperature will deviate from the sensor temperature as a result of thermal lags that become increasingly pronounced at high heating rates [19,20]. The continuous and automatic adjustment of the heater power necessary to keep the sample holder temperature identical to that of

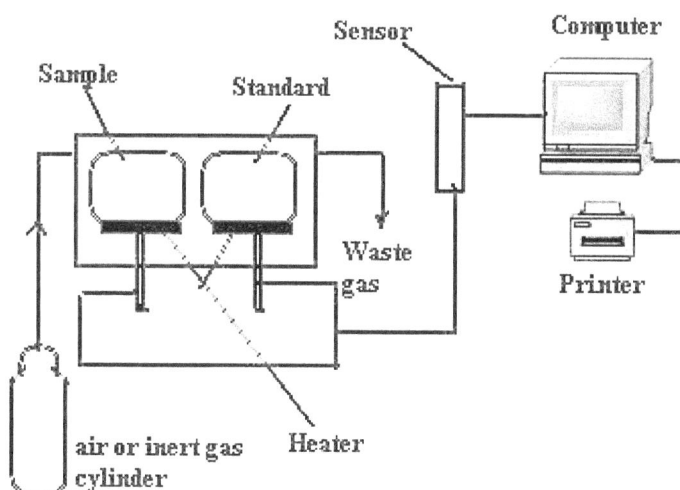

Figure 4.2.   Diagram of a typical differential scanning calorimeter.

the reference holder, that is, to keep $\Delta T = 0$, provides a varying electrical signal that is opposite but equivalent to the power absorbed or released by the sample. Thus, the system is a power-compensated DSC.

4.2.3.   Advantages

- DSC measurements are used only within narrow ranges of temperature in order to record the phase transitions such as melting and not observing the thermal effects of degradation at high temperature [21–28].
- Enthalpy changes can be detected and obtained with DSC.
- It is a fast screening method for new materials.
- It is a tool that is very much used in research.

4.2.4.   Disadvantage

- The sample will be destroyed and only final ash remains: It is a destructive test.

4.3.   Differential Thermal Analysis

The technique of differential thermal analysis (DTA) has been applied to many problems related to the melting behavior of polymers. Melting points are usually derived from the endothermic maximum of the fusion curve, crystallinities from its area, and the crystallite size distribution qualitatively inferred from its overall breadth and shape [21–23]. However, due to the influence of kinetic factors in bulk crystallization, the temperature–crystallinity dependence of a semicrystalline polymer is not uniquely determined by its component structures and may be varied over a wide range according to the thermal pretreatment of the sample.

Traditionally, either microscopic and/or dilatometric techniques were used to examine the first- and second-order phase transitions. Instrumental methods replace the traditional techniques to examine the phase transitions. Without quantitative calorimetric information, DTA, which is a sensitive temperature differential, reflects the heat capacity (C_p) and thermal events that take place during the temperature scan.

DTA is very useful for characterization of the polymer melting process and the thermal behavior of polymers. In many instances the determination of temperatures of transformation and crystallization is possible by means of DTA/DSC. Thermal methods, especially DTA, DSC, TGA, and dilatometry, are widely used for the characterization of polymers. There are two main fields of application of these methods:

1. Determination of the characteristic glass points and of thermodynamic properties

2. Characterization of the melting process and the thermal behavior of polymers and melts

Because the properties of polymers are dependent on their thermal history, the polymers are considered in terms of the temperature and time required to form and to analyze them. This means that the cooling and heating rates must be stated, and also details of the sample preparation, sample weight, and instrumentation [24].

Differential thermal analysis is a technique developed for studying the phenomena occurring when materials are heated. DTA characterizes the polymer molecular structures in terms of first- and second-order transitions. Differential thermal analysis with heating, cooling or heating, and isothermal scans in the technological developments helps to measure specimens in comparison with standards. It measures the temperature difference between a specimen and a standard during the entire system by heating or cooling under a controlled heating or cooling rate. The results give the information on the first- and second-phase transitions. DTA is an indispensible tool for dealing particularly with semicrystalline polymers.

4.3.1. Principle

Differential thermal analysis measures the thermal effects occurring in the sample by continuously recording the temperature difference between the sample and a reference material as a function of the sample temperature. Differential thermal analysis has been used for many years for detecting phase transitions.

Application of DTA to polymers has increased. DTA is used to determine glass transition temperatures as a function of polymer molecular weight. DTA is said to be at least as accurate as other techniques. It is far more rapidly conducted. Very good correlation has also been found between DTA and thermogravimetric analysis. Reactions such as the evolution of volatiles, polymerization, and condensation are recordable, and reproducibly so.

4.3.2.   Instrumentation

Practice involves charging a material into a platinum crucible and following the response of a single, centrally embedded thermocouple upon heating. This technique has been advanced to a single recording which gives results from two galvanometers, one for the material and the other for a reference substance. Differential thermal analysis devised the differential thermocouple circuit from which the process derives its name. The differential thermocouple is used to measure the difference in temperature between the sample and ambient furnace condition. The temperature of a reference material is used rather than the furnace temperature for comparison.

Sample holders have ranged from inexpensive and simple to complex and costly. The thermal behavior of substances in the liquid state can be followed in other types of systems by using an inert material as an absorbent. This prevents the material from running out of the sample holder.

Many varieties of thermocouples have been used in differential thermal analyzers. They fall into two general classes, base metal and noble metal. The choice of thermocouples for a DTA apparatus is governed by the temperature limitations to be imposed, the thermal response desired from the thermocouple, and the chemical reactivity of the materials to be investigated. The base-metal thermocouples have to be applied with more caution because of their reactivity. They have much greater differential electromotive force response to thermal excitation than noble-metal types, which are less expensive. Opposing physical and electrical factors are encountered in thermocouple circuits.

4.4.   Polymer Degradation

Thermal analysis is important in polymer degradation. Degradation of polymers is essential to analyze both thermogravimetric and differential calorimetric data over a large temperature range at similar heating rates and under similar other conditions. Studies on thermo-oxidative and thermal stability of polymeric materials are available by simultaneous measurements by DSC and TGA methods over a wide range of temperature.

The deterioration of the physical, chemical, and/or aesthetic properties of polymers which may occur during processing or subsequent usage has been the subject of long-standing interest and concern to polymer producers.

The polymer degradation chemistry is not describing complicated kinetics of decomposition at its current status. During processing at defined shear rates with temperature in the presence of oxygen, mechanical initiation and thermal oxidation transformations occur. Knowledge of the degradation factors and degradation mechanisms is required in order to develop appropriate procedures that permit an evaluation of the properties of the material.

The thermal degradation of samples can be divided into three stages:

1. The drying stage, where the volatiles are evolved
2. The main pyrolysis stage, where the chemical decomposition occurs
3. The carbonization stage, where the total decomposition occurs and ends with residue.

4.4.1. Thermal Degradation Kinetics

In study of the degradation of polymers, account must be taken of the effect of the molecular-weight distribution (MWD) of the polymer on the nature of the process and on the change in MWD during the course of degradation. Failure to take account of the MWD can lead to serious errors in the final conclusions about the mechanism of the process. Conversely, theoretical and experimental study of the effect of MWD on the characteristics of the process and on the change in MWD or weight-average and number-average molecular weights can provide an additional source of information on the degradation process. From what follows, it will also be evident that study of the kinetics of degradation and of the change in the average molecular weights during degradation can give a method of analysis of the MWD of polymers [25].

Kinetic analysis of thermal decomposition processes has been a subject of interest for many investigators throughout the modern history of thermal decomposition. The interest is fully justified. On one hand, kinetic data are essential for designing any kind of device in which thermal decomposition takes place; on the other hand, kinetics is intrinsically related to the decomposition mechanisms. Knowledge of the mechanism allows the postulation of kinetic equations or vice versa, and kinetics is the starting point to postulate mechanisms for the thermal decomposition [26].

Although kinetic studies can be performed in different devices, thermogravimetry (TG) is, by and large, the most used technique. This technique consists of preheating the sample to a given temperature and then starting the experiment with a fixed nominal heating rate.

The kinetic analysis provides information on the energy barriers of the process and clues to the degradation mechanism for polymers. The challenge for the study of thermal degradation kinetics is to find a reliable approach.

Because single-heating-rate methods have identified shortcomings [27], reliable multiheating-rate methods have been used extensively to study the thermal degradation kinetics for polymers [28,29].

Thermogravimetric analysis (TGA) is usually used to elucidate the polymer degradation [30–36]. A loss of mass is measured, and the polymeric materials degradation frequently starts with enthalpy changes, particularly true for thermo-oxidation. In TGA, thermal effects are observed separately but in parallel with the mass loss over a wide range of temperature and can be ascribed precisely to specific mechanisms of degradation. In isothermal degradation studies, change in the weight of the sample is recorded as a function of time.

References

1. L. M. Clarebrough, M. E. Hargreaves, D. Michell, and G. W. West, *Proc. Roy. Soc. Lond. A* 125, 507 (1952).
2. J. Chiu, *Anal.Chem.* 34, 1841(1962).
3. S. B. Warrington, in H. Günzler and A. Willimas (eds.), *Handbook of Analytical Techniques,* p. 83344, Wiley-VCH Verlag, Weinheim, Germany (2001).
4. S. B. Warrington, in E. W. Charsley and S. B. Warrington (eds.), *Thermal Analysis—Techniques and Applications,* Royal Society of Chemistry, London, (1992).
5. M. R. Holdiness, *Thermochim. Acta* 75, 361 (1984).
6. G. Szekely, M. Nebuloni, and L. F. Zerilli, *Thermochim. Acta* 196, 511 (1992).
7. A. P. Snyder, A. Tripathi, J. P. Dworzanski, W. M. Maswadeh, and C. H. Wick, *Anal. Chim. Acta* 536, 283 (2005).
8. A. Fernández, J. Torrecilla, J. Garćia, and F. Rodríguez, *J. Chem. Eng. Data* 52, 1979 (2007).
9. J. Chiu, *Appl. Polym. Symp.* 2, 25 (1966).
10. R. G. Craig, J. M. Powers and F. A. Peyton, *J. Dent. Res.* 50, 450 (1971).
11. W. W. Wendlandt, *Thermal Methods of Analysis,* Academic Press, New York (1965).
12. E. L. Simons, A. E. Newkirk, and I. Aliferis, *Anal. Chem.* 29, 48 (1957).
13. J. Rouquerol, *Thermochim. Acta* 144, 209 (1989).
14. R. Shutt, C. Turmel, and B. Touzard, in *Proceedings of the 5th Eurobitume Congress,* Stockholm, 1A(1.11), 76 (1993).
15. B. Brule and S. Gazeau S., in *Proceedings of the ACS Symposium on Modified Asphalts,* Orlando, Florida (1996).
16. H. J. Borchardt and F. Daniels, *J. Am. Chem. Soc.* 79, 41 (1957).
17. H. E. Kissinoer, *Anal. Chem.* 29, 1702 (1957).
18. M. Gordon and W. Simpson, *Polymer* 2, 383 (1961).
19. A. L. Greer, *Acta Metall.* 30, 171 (1982).

20. G. W. H. Hohne and E. Gloggler, *Thermochim. Acta* 151, 295 (1989).
21. B. Ke, *J. Polym. Sci.* 42, 15 (1960).
22. B. Ke, *J. Polym. Sci.* 61, 47 (1962).
23. B. Wunderlich and W. H. Kashdan, *J. Polym. Sci.* 50, 71 (1961).
24. K. Heide, *Thermochim. Acta* 110, 419 (1987).
25. A. A. Berlin and N. S. Yenikolopyan, *Vysokomol. Soyed.* A10: No. 7, 1475, (1968).
26. J. A. Conesa, A. Marcilla, J. A. Caballero, and R. Font, *J. Anal. Appl. Pyrol.* 58/59, 617 (2001).
27. S. Vyazovkin and N,. Sbirrazzuoli, *Macromol. Chem. Phys.* 200, 2294 (1999).
28. S. Vyazovkin, J. Comput. Chem. 22, 178 (2001).
29. M. E. Brown, M. Maciejewski, S. Vyazovkin, R. Nomen, J. Sempere, A. Burnham, et al., *Thermochim. Acta* 355, 125 (2000).
30. J. Ma, Z. Qi, and Y. Hu, *J. Appl. Polym. Sci.* 82, 3611 (2001).
31. M. Zanetti, G. Camino, P. Reichert, and R. Mulhaupt, *Macromol. Rapid Commun.* 22, 176 (2001).
32. Y. Tang, Y. Hu, L. Song, R. W. Zong, Z. Gui, Z. Y, Chen, et al., *Polym. Degrad. Stabil.* 82, 127 (2003).
33. H, Qin, S. Zhang, C. Zhao, M. Feng, M. Yang, Z. Shu, et al., *Polym. Degrad. Stabil.* 85, 807 (2004).
34. J. G. Zhang, D. D. Jiang, and C. A. Wilkie, *Thermochim. Acta* 430, 107 (2005).
35. S. U. Lee, I. H. Oh, J. H. Lee, K. Y. Choi, and S. G. Lee, *Polym. Korea* 29, 271 (2005).
36. W. Gianelli, G. Ferrara, G. Camino, G. Pellegatti, J. Rosenthal, and R. C. Trombini, *Polymer* 46, 7037 (2005).

Chapter 5

Rheology and Other Instrumental Techniques

Instrumental methods are among the most powerful techniques for polymer analysis. Because of their simplicity, sensitivity, versatility, and speed, their applications in polymer analysis are growing at a rapid pace. Polymer testing has come a long way from the determination of molecular weight with respect to molecular properties used for in-situ monitoring of rheological behavior.

Rheological testing is often necessary to assess processing characteristics. Rheology is widely used to provide information about the molecular structure and composition of polymeric systems [1]. Characteristics of interest include molecular weight, molecular-weight distribution, level and type of long-chain branching, and the state of dispersion. Most characterization techniques involve the measurement of rheological properties of the melt; however, viscosity measurement by dilute solution viscosity is used to track molecular weight when it is not convenient to work with the melt.

Rheology is of central importance in polymer processing. During processing, the flow details and pressure drop are governed essentially by the viscous (shear) properties of the polymer melt. When the melt leaves the die, the flow changes in nature and becomes extensional. The flow behavior in response to these forces, including both shear and extensional flow properties, is governed entirely by the rheological properties of the melt. Moreover, because of the memory effects of polymer melts, the processing behavior also depends on the deformation history experienced by the melt while in the processing equipment. Therefore, both shear and extensional rheological

DOI: 10.5643/9781606502440/ch5

properties of the polymers can be expected to play significant roles in the processing [2–6].

5.1. Rheology

Rheology is the study of the deformation and flow behavior of materials. An operational definition is "the study of the response of certain materials to the stresses imposed on them."[7] It is also concerned with establishing predictions for mechanical behavior (on a continuum mechanical scale) based on the micro- or nanostructure of the material [8].

Rheology seeks to understand the relationship between applied force, or stress, and the resulting deformation, particularly for materials that show nonsimple responses. Applied rheology endeavors to connect fundamental properties and real processes. This chapter will very briefly highlight some key topics and newer approaches to the rheological analysis of coatings, with an eye to understanding coating flows and other pertinent issues.

Rheological characteristics may be important during four stages of processing: before mixing [9–11], during mixing, during setting [12–16], and after the completion of setting [17–19].

The rheological properties of polymers are affected by a number of factors, with molecular structure being one of the most important [20,21]. Here, molecular structure refers to molecular-weight distribution (MWD) and the parameters that describe it: number-, weight-, and z-average molecular weight (Mn, Mw, and Mz, respectively); polydispersity index (PI), defined as the ratio of Mw to Mn; and the degree of branching (short- and long-chain branching). Rheological evaluation of polymer products is a useful tool in characterizing materials for the purpose of process quality control. For thermoplastic polymers, dilute solution viscosity and flow rate are probably the most important parameters.

5.1.1. Principle

Shear depends not only on melt viscosity, but is also affected by the molecular properties of the polymer, including MWD, branching, etc. Polymer characterization simulates polymer processing techniques such as extrusion, injection, etc. Hence the melt viscosity, time, and temperature of a polymer process all depend at the molecular level on the shear rate. In addition to melt index and Mooney viscosity, measurement of rheological properties is one of the workhorses used for simple evaluation of polymer processing.

5.1.2.   The Rheometer

Processing history has a significant effect on the rheological properties of branched polymers, especially on the melt elasticity and the rheological properties during elongational flow, because entanglement couplings associated with long chain branches are primarily responsible for these properties.

A capillary rheometer is used to determine viscosity at high shear rates because it reflects polymer processing behavior well. The rheometer can estimate MWD even for polymers for which there is no suitable solvent analysis of solution properties. Melt strength, defined as the force needed for extension of a polymer extrudate, is measured using a capillary rheometer.

Parallel-plate rheometry is often performed to determine the linear viscoelastic properties of pure components and their blends. The measurements are performed using a controlled-stress rheometer at various temperatures, usually 130, 150, 170, 190, and 210°C. Master curves have been obtained using the time–temperature superposition principle (TTS) [22,23].

5.1.3.   Instrumentation

Quality control is a special type of characterization in which the objective is to ensure that a manufacturing process is under statistical control and that the product remains suitable for its intended end use over an entire production run [24]. Obviously, measurement of the viscosity at a number of shear rates at several frequencies can be used to detect variations in molecular structure, and such tests are often used for quality control. However, a more common procedure is to use a simpler, one-point empirical test method for a well-defined rheological property. This has the advantage that the apparatus required is relatively inexpensive and easy to use. On the other hand, such a test may fail to reveal a shift in the shape of the molecular-weight distribution. The most widely used empirical test is the measurement of the flow rate, or melt index, using an extrusion parameter, more commonly known as a melt indexer. It is now well known that at high stress levels, the assumption that a melt is a continuum fails, and constitutive equations are no longer valid under such conditions [25]. The phenomena of gross melt fracture, sharkskin, and wall slip occur in the flow of most linear high-molecular-weight polymers, but the understanding of these phenomena is very limited. In particular, no models are available to predict when they will occur and how they affect the overall flow behavior of a melt. This is a difficult area of research, but it is essential that it be addressed if progress in plastics processing is to be made in an orderly way.

The shear cell used in rheometer measurements contains sample placed between two parallel glass plates [26]. The bottom plate remains stationary while the top one rotates at a user-controlled speed. The whole assembly is placed inside a brass block which can be heated or cooled continuously, or maintained at a constant temperature. The bottom part of the cell stands on a freely rotating shaft with an extended arm resting on a calibrated spring. When shear is applied to the top plate, the lower part of the cell rotates as well. The displacement of the arm can be measured by means of a displacement transducer which gives values for the torque, and hence the viscosity and stress can be obtained. At the end of an observation the plate/sample/plate assembly can be quickly released and quenched for further measurements.

5.1.4.   The Online Rheometer

The concept of the online rotational rheometer is shown in Figure 5.1 [27]. The melt is collected within the extruder by opening the rotating tap and shaped as required, typically into a disk or a tapered disk for parallel plates and cone-and-plate measuring geometries, respectively. The melt is squeezed in between the lower and upper plates. Measurements are taken using the

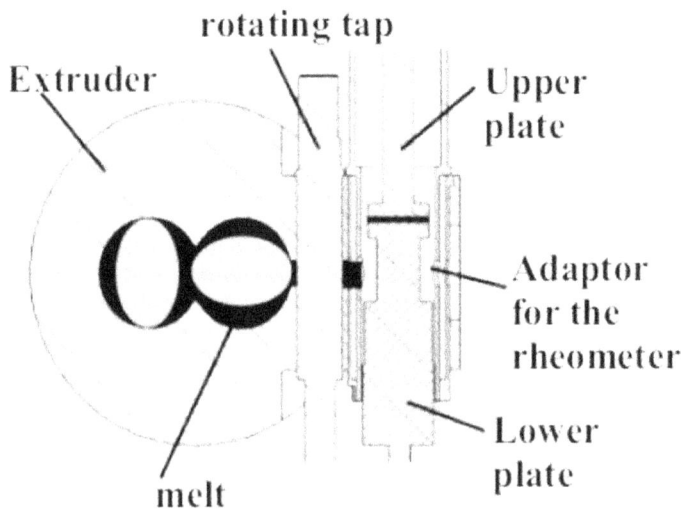

Figure 5.1.   The online rheometer concept. [Reprinted with permission from S. Mould, J. Barbas, A. V. Machado, J. M. Nóbrega, and J. A. Covas, *Polym. Testing* 30, 602–610 (2011). Copyright © 2011 Elsevier Ltd. All rights reserved.]

motion-sensing and controlling capacities of a commercial rotational rheometer. The commercial rotational rheometer is coupled to the upper plate and fixed to the device by means of an adaptor. As in conventional offline rheometry, the instrument can operate in either steady or oscillatory mode. A variety of testing is possible, including measurement of isothermal steady shear, frequency sweep, amplitude sweep, time sweep, and step stress/strain (creep and recovery/stress relaxation) [28].

5.1.5. Advantages

- A major application of polymer melt rheology is the characterization of products, in both development and production.
- It is useful in testing new types of catalysts and polymerization or degradation processes, finding new additives and blends, as well as in quality control—wherever there is a need for sensitive methods to characterize materials.

5.1.6. Disadvantages

- Rheology can be used as an online method and is extremely sensitive in the high-molecular-mass range, but it does not yield the molar mass distribution (MMD) or information about branching directly.

5.2. Mass Spectrometry

In the past, mass spectrometry (MS) of synthetic polymers was hardly possible; as a rule, polymers had to be degraded thermally or chemically prior to mass spectroscopic analysis. For many years, field desorption (FD) on special instruments was the only option for mass spectrometric analysis of intact polymers up to 10 kDa in size [29]. Today, however, among the analytical techniques used in the characterization of synthetic polymers, mass spectrometry is of increasing importance [30].

Mass spectrometers are often connected physically or electronically to other instruments as well as to a computer. These are typically not individual instruments. Infrared (IR) or nuclear magnetic resonance (NMR) spectrometry can identify compounds with specificity comparable to that of mass spectrometry, and MS does not absorb radiation such as infrared, ultraviolet, or radio waves. However, MS is becoming a viable alternative to more traditional methods for polymer analysis because it has the ability to provide absolute masses.

Mass spectrometry has been used extensively to analyze additives or impurities in polymers and to obtain ion patterns by pyrolysis or thermal degradation for polymer identification [31–33]. Most recently, ionization methods have been developed that make characterization of intact polymers possible. Even with the wide array of ionization methods potentially available for mass analysis of intact polymers, mass spectrometry has seldom been used as the definitive method to determine molecular weights and molecular-weight distributions of polymers. Many polymers are analyzed by time-of-flight methods because of the extended mass ranges afforded by mass spectrometers.

5.2.1.   Principle

Polymers are mixtures of discrete compounds that differ in numbers and types of repeat units, end groups, architecture, and so forth, and accurate molecular-weight analysis requires not only mass accuracy for each oligomer, but the absence of fragmentation and a signal response that is independent of oligomer mass or at least a known function of mass.

The material under test is ionized in the mass spectrometer. Using a beam of high-energy electrons, the electrically neutral molecules are converted to molecular ions. Fragmentation immediately follows the ionization. The bonds break, and in many instances new bonds form based on the structural characteristics of the fragmenting ion.

5.2.2.   Advantages

- Mass spectrometry has gained wide acceptance as a tool for polymer characterization because of the precision of the data, which can be used to relate the sharp peaks to the molecular structure.

5.2.3.   Disadvantages

- Sample preparation is less straightforward due to the coexistence of several distributions, and homogeneous co-crystallization between the matrix and the synthetic polymer is not easily achieved.
- MS is quantitatively much less reliable than chromatography. Apart from different manners of data presentation, mass discrimination

occurs during ionization, transmission, and detection of wide polymer distributions.

5.3.   Matrix-Assisted Laser Desorption Ionization (MALDI) Mass Spectrometry

The introduction of matrix-assisted laser desorption ionization (MALDI) [34,35] and electrospray ionization (ESI) [36] has enabled the formation of gas-phase ions from a wide variety of synthetic polymers, opening a new era for their mass spectrometric analysis. MS experiments provide the mass-to-charge ratios (*m/z*) of the constituent *n*-mers of a polymeric material, from which compositional heterogeneity, molecular weight, and functionality distributions can be deduced. Such information has been essential in the discovery of new polymerization techniques, the elucidation of polymerization mechanisms, and the advancement and commercialization of new products [37–47]. Modern soft ionization techniques such as electrospray ionization (ESI) [48] and MALDI [49,50] have had and still have an enormous impact on the analysis of polymers.

MALDI/time-of-flight (MALDI-TOF) mass spectrometry can reveal the absolute molecular weight of each discrete macromolecule within the polymer distribution. This, in turn, allows for much more detailed analysis, including determination of the exact mass of the repeat units, as well as the end-group masses. MALDI-TOF MS offers a unique advantage in that all of the signals observed contain information about the masses of the end groups present, in contrast to the small fraction of the NMR signal that contains information about the end groups.

Modern MALDI-TOF mass spectrometry is a powerful technique for the fast and accurate determination of a variety of polymer characteristics. The determination of absolute molecular weights of individual polymer chains provides much information, such as the repeat unit and the kinds of end groups.

5.3.1.   Sample Preparation

There are several methods of sample preparation, but the most commonly used is the dry droplet method as shown in Figure 5.2 [51]. The polymer sample is dissolved in a suitable solvent andv mixed with an excess of small UV-absorbing molecules (the matrix) dissolved in the appropriate mixture of organic solvent/water. The solvent is then evaporated, leaving matrix crystals containing the analyte [52]. Both the analyte and the matrix molecules

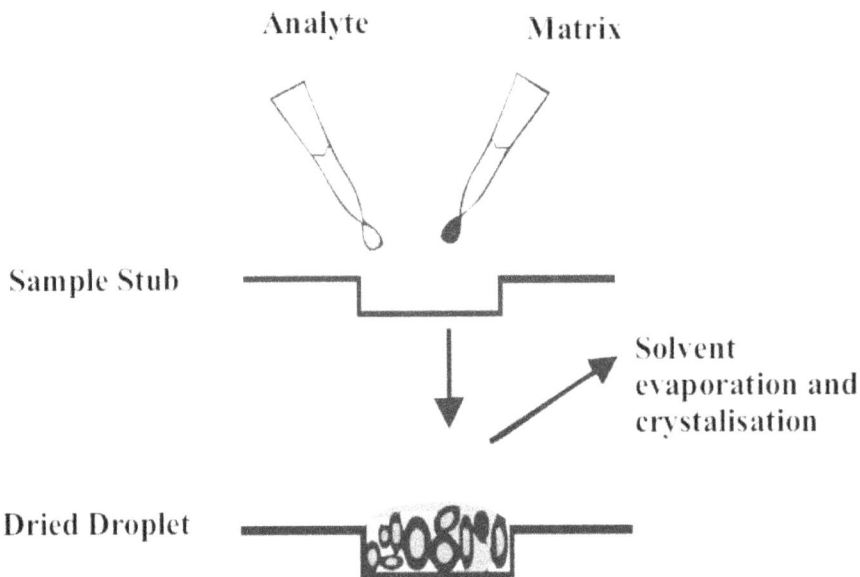

Figure 5.2. Sample preparation for conventional MALDI-MS. [Reprinted with permission from H. J. Griesser, P. Kingshott, S. L. McArthur, K. M. McLean, G. R. Kinsel, and R. B. Timmons, *Biomaterials* 25, 4861–4875 (2004) . Copyright © 2004 Elsevier Ltd. All rights reserved.]

must be dissolvable under the same conditions to prevent preferential pre-cipitation of one of the components [53]. The sample is then introduced to the spectrometer and the dried crystals are irradiated with a pulsed laser which achieves rapid evaporative heating of the crystals.

5.3.2.   Instrumentation

MALDI-TOF MS is ideally suited for polymer analysis because of the simplicity of the mass spectra, which show mainly single-charged quasi-molecular ions with hardly any fragmentation; the time-of-flight (TOF) ana-lyzer, in which very-high-molecular-weight polymers can be analyzed; and the state-of-the art reflection instruments equipped with a delayed-extrac-tion ion source [54,55], which offer a higher resolution range and thereby allow—depending on the monomer mass and the polymer complexity—the determination of the repeating mass increment (i.e., the monomer mass) and the mass residue (i.e., the end groups plus known cation mass) up to 35 kDa. Figure 5.3 [56] shows a diagram of a TOF/TOF mass spectrometer showing

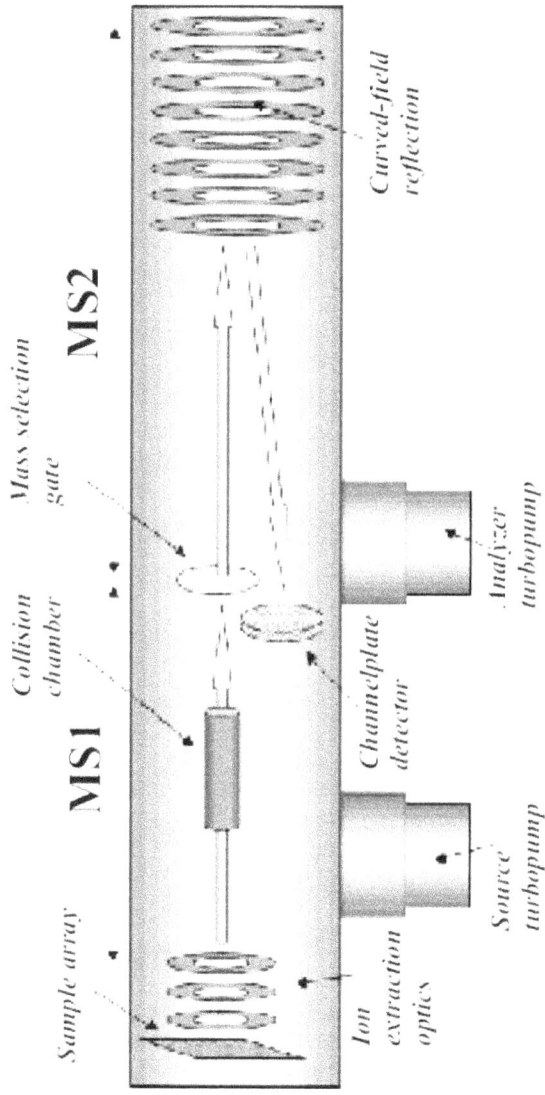

Figure 5.3. Diagram of a TOF/TOF mass spectrometer showing the location of the collection chamber, the mass selection point for the first mass analyzer, and the second mass analyzer with its curved field reflection. [Reprinted with permission from R. J. Cotter, B. D. Gardner, S. Litchenko, and R. D. English, *Anal. Chem. 76*, 1976 (2004). Copyright © 2004, American Chemical Society.]

the location of the collection chamber, the mass selection point for the first analyzer, and the second mass analyzer with its curved field reflection.

5.3.3. Advantages

- MALDI-TOF mass spectrometry is particularly applicable to the measurement of polymer distributions, because it mainly produces singly charged species, which in general can be interpreted easily.
- It allows for the measurement of extremely high molecular weights with virtually no fragmentation [57].
- In a mass range where single polymer chains are resolved, MALDI-TOF enables determination of repeating units and end-group compositions [58–60].
- In the region below 20 kDa, where other absolute methods such as osmometric mass analysis might give inaccurate results, MALDI-TOF mass spectrometry can be used as a supplemental, independent, absolute method.

5.3.4. Disadvantages

- Polymer characterization by MALDI-TOF MS has limitations for quantitative analysis because of differences in ionization efficiency with respect to differing molecular weights or end-group functionalities.
- MALDI-TOF MS remains a powerful tool for polymer characterization, but it must be used in concert with other analytical techniques to provide a realistic analysis of a polymer sample.

5.4. Electron Microscopy

Traditional polymer microscopy is based largely on differential staining to resolve chemical information in a multiphase polymer microstructure indirectly. Although traditional techniques of electron microscopy involving heavy-element stains and bright-field imaging have certainly made, and will continue to make, a significant contribution to the understanding of polymer microstructure, these techniques are unable to address many current and emerging problems. Electron microscopy techniques provide more resolvable detail than optical microscopy. The major historical contribution of electron microscopy and electron diffraction to the understanding of polymer structure has been the introduction of the concept of chain folding.

Polymer single crystals are ideally suited for investigations by electron microscopy and electron diffraction because of their limited thickness (10 nm or tens of nanometers). However, one of the major limitations is the sensitivity of the electron beam, which requires working with extremely low electron doses and, usually, nonconventional electron microscope settings. Major insights into the structure of polymers have been gained through investigation of polymer single crystals, and with both components of the single crystals: the lamellar crystalline core, and the less ordered layers made of folds that sandwich all polymer lamellar crystals. Indeed, contrary to most or all other materials, polymer single crystals are actually composites, with a molecularly connected crystalline core and two amorphous surfaces.

Polymer single crystals have been used to investigate many aspects of the crystal structure. In many cases, polymers display structural polymorphism, with many polymorphs being unstable to, e.g., mechanical deformation. Investigation of these crystal structures cannot therefore rely on the standard x-ray fiber diffraction analysis.

5.4.1. Scanning Electron Microscopy (SEM)

Scanning electron microscopy fracture analysis continues to be the best method for assessing structure–property relations, especially for toughness. SEM resolution is traditionally associated with the electron-beam spot size, which is smaller at higher acceleration voltage (AV). For high-resolution microscopy of polymers, however, resolution and information-passing capacity [61] are inherently linked to the sample volume, where secondary electrons are created and from which they are collected, and also to the efficiency of the low-energy electron detectors. The contrast/brightness values are also a function of the electron flux per unit time, and discharge mechanisms.

Scanning electron microscopy has shown promise in the field for general morphology characterization. However, with the advent of a new generation of high-resolution SEM, it is now possible not only to determine the general phase morphology, but also the lamellar structure [62]. There are several problems, however, in relying exclusively on block face imaging methods. Some polymer materials require heavy-metal staining of thin sections to reveal the detailed morphology, because no good method of block staining has yet been identified. On the other hand, due to the reduced beam spread as a focused beam passes through a thin section as compared to a bulk sample, higher-spatial-resolution microanalysis can be achieved.

One of the biggest problems with SEM of polymers at conventional voltages is their great propensity to accumulate negative charge and consequently to reject the incident probe in successive scans. This occurs, of course, because the electrons that impinge on the sample surface do not drain away faster than they are supplied by an intense, high-energy probe. The physical attribute of conductivity allows accumulated charge to be dispersed in the case of metal and some semiconductor samples, but this is not an advantage enjoyed by polymers and other nonconductors. However, conductivity is not the only way that materials disperse accumulated charge from an incident probe: They also emit secondary (SE) and back-scattered (BS) electrons. At very low accelerating voltages the fact that the SE escape depth is of the same order as the incident electron penetration depth can be used to eliminate the buildup of surface electron charge.

SEM provides a useful tool for the evaluation of polymers and their end products during manufacturing processes, and has proved to be a very useful instrument for the assessment of polymer morphology. The three-dimensional images produced by SEM clearly show surface features, such as the presence of surface modifications, finish applications, wear, and the nature and cause of polymer failure. The SEM is thus useful for evaluating construction, coverage, uniformity, surface structure, and effects of wear.

Microstructure studies generally require complementary optical and SEM study to understand the arrangement of the fine structural details within the macrostructure.

5.4.2. Transmission Electron Microscopy (TEM)

Transmission electron microscopy is an important instrumental method used to understand better the organization of functional polymer systems. TEM has been used extensively ever since its invention. With the further development of electron tomography, detailed analysis of 3-D volume organization has become available with nanometer resolution. The results from TEM image analysis show the distribution and fundamental structure–property correlations. It is useful for both fundamental and applied analysis of promising polymeric materials.

In the evolving field of microscopy, a wide range of instruments can be utilized to resolve details in crystalline and semicrystalline polymers ranging from the millimeter to subnanometer size scales. Light microscopy can be utilized for examination of spherulitic structures, but more detailed examination of single lamellae crystals and polymer interfacial interactions have relied heavily on TEM.

The increased replacement of metals and ceramics by polymeric materials in automotive, consumer goods, and medical devices has resulted in a demand for faster methods to study the microstructure of these materials. Traditionally, microstructural analysis of polymeric materials has been carried out using TEM. However, specimen preparation for TEM analysis is tedious and skill-intensive, and TEM is not generally a viable option for polymeric materials containing inorganic fillers, due to ultramicrotomy limitations. Since the 1990s, alternative methods, based on preferential etching of phases in polymer blends by exposure to plasma or solvents [63], have been employed to generate the necessary topographical contrast, followed by imaging using scanning electron microscopy. These techniques have greatly simplified specimen preparation and have resulted in faster analysis and increased productivity.

In addition to probing long-term structural changes, methods are available for probing transient structural changes which occur during shear and immediately upon cessation of flow [64,65]. Traditional parallel superpositions have been been used to investigate microstructural changes in liquid crystalline polymers [66]. The superposition provides a transient probe which may be used to investigate structural changes that occur as the shearing deformation is taking place.

Transmission electron microscopy techniques are important for the elucidation of details of microstructure. The types of detailed structures that can be determined by TEM include:

- Polymer structure
- Void size, shape, and distribution
- Size, shape, and distribution of fillers
- Local crystallinity
- Crystallite sizes

5.5. Future Trends

The instrumental methods are very powerful in providing important information about polymers, both before and during processing, especially in such applications as coatings, composite materials, biomaterials, etc. They are widely used to aid in modifying and optimizing polymers for these and other applications.

References

1. J. M. Dealy and K. F. Wissbrun, *Melt Rheology and Its Role in Plastics Processing,* corrected paperback edition, Kluwer Academic Publishers, Dordrecht, The Netherlands (1999).
2. J. M. Dealy and K. F. Wissbrun, *Melt Rheology and Its Role in Plastics Processing,* Reinhold, New York (1995).
3. D. V. Rosato and D. V. Rosato (eds.), *Blow Molding Handbook,* Hanser, New York (1989).
4. E. D. Henze and W. C. L. Wu, *Polym. Eng. Sci.* 13, 153 (1973).
5. D. M. Kaylon and M. R. Kamal, *Polym. Eng. Sci.* 26, 508 (1986).
6. D. W. Van Krevelen, *Properties of Polymers: Their Correlation with Chemical Structure; Their Numerical Estimation and Prediction from Additive Group Contributions,* Elsevier, New York (1990).
7. R. R. Eley, in *ASTM Paint and Coatings Testing Manual,* 14th ed., ASTM, Philadelphia, pp. 333–368 (1995).
8. P. Prentice, *Rheology and Its Role in Plastics Processing* (Rapra review report), Rapra Technology Ltd., Shrewsbury, U.K., vol. 7, p. 3 (1995).
9. M. Braden, *J. Dent. Res.* 46, 429 (1967).
10. E. C. Combe and J. B. Moser, *J. Dent. Res.* 57, 221 (1978).
11. T. W. Herfort, W. W. Gerberich, C. W. Macosko, and R. J. Goodkind, *J. Prosthet. Dent.* 38, 396 (1977).
12. H. J. Wilson, *Br. Dent. J.* 121, 277 (1966).
13. M. Braden, *J. Dent. Res.* 45, 1065 (1966).
14. W. D. Cook, *Biomed. Mater. Res.* 16, 315 (1982).
15. W. D. COOK, *Biomed. Mater. Res.* 16, 345 (1982).
16. J. F. McCabe and A. J. Bowman, *Br. Dent. J.* 151, 179 (1981).
17. H. J. Wilson, *Br. Dent. J.* 121, 322 (1966).
18. A. J. Goldberg, *J. Dent. Res.* 53, 1033 (1974).
19. N. S. Salem, D. C. Watts, and E. C. Combe, *Dent. Mater.* 3, 37 (1987).
20. J. M. Dealy and K. F. Wissbrun, *Melt Rheology and Its Role in Plastics Processing,* Reinhold, New York (1995).
21. D. W. Van Krevelen, *Properties of Polymers: Their Correlation with Chemical Structure; Their Numerical Estimation and Prediction from Additive Group Contributions,* Elsevier, New York (1990).
22. M. L. Williams, R. F. Landel, and J. D. Ferry, *J. Am. Chem. Soc.* 77, 3701 (1955).
23. J. van Gurp, and J. Palmen, in *Proc. 12th Int. Congress on Rheology,* Quebec City, Canada, p. 134 (1996).
24. J. M. Dealy and P. C. Saucier, *Rheology as a Tool for Quality Control in the Plastics Industry,* Hanser-Gardner, Cincinnati, OH (1999).
25. J. M. Dealy, in *Progress and Trends in Rheology V (Proc. Fifth European Rheology Conf.),* p. 8 (1998).
26. I. A. Hindawi, J. S. Higgins, and R. A. Weiss, *Polymer* 33, 2522 (1992).
27. S. Mould, J. Barbas, A. V. Machado, J. M. Nóbrega, and J. A. Covas, *Polym. Testing* 30, 602 (2011).

28. J. A. Covas, J. M. Maia, A. V. Machado, and P. Costa, *J. Non-Newtonian Fluid Mech.* 148, 88 (2008).
29. L. Prokai, *Field Desorption Mass Spectrometry,* Marcel Dekker, New York (1990).
30. P. B. Smith, A. J. Pasztor, Jr., M. L. McKelvy, D. M. Meunier, S. W. Froelicher, and F. C.-Y. Wang, *Anal. Chem.* 69, 95R (1997).
31. P. Vouros and J. W. Wronka, in H. G. Barth and J. W. Mays (eds.), *Modern Methods of Polymer Characterization,* chap. 12, John Wiley, New Delhi (1991).
32. K. D. Cook, in *Encyclopedia of Polymer Science and Engineering,* 2nd ed., Vol. 9, p. 319, John Wiley, New York (1987).
33. H.-R. Shulten and R. P. Lattimer, *Mass Spectrum. Rev.,* 3, 231 (1984).
34. M. Karas and F. Hillenkamp, *Anal. Chem.* 60, 2299 (1988).
35. K. Tanaka, H. Waki, S. Ido, Y. Akita, Y. Yoshida, and T. Yoshida, *Rapid Commun. Mass Spectrom.* 2, 151 (1988).
36. J. B. Fenn, M. Mann, C. K. Meng, S. F. Wong, and C. M. Whitehouse, *Science* 246, 64 (1989).
37. S. D. Hanton, *Chem. Rev.* 101, 527 (2001).
38. G. Montaudo and R. P. Lattimer (eds.), *Mass Spectrometry of Polymers,* CRC Press, Boca Raton, FL (2002).
39. C. N. McEwen and P. M. Peacock, *Anal. Chem.* 74, 2743 (2002).
40. R. Murgasova and D. M. Hercules, *Anal. Bioanal. Chem.* 373, 481 (2002).
41. H. Pasch and W. Schrepp, *MALDI-TOF Mass Spectrometry of Synthetic Polymers,* Springer-Verlag, Berlin (2003).
42. P. M. Peacock and C. N. McEwen, *Anal. Chem.* 76, 3417 (2004).
43. M. A. Arnould, M. J. Polce, R. P. Quirk, and C. Wesdemiotis, *Int. J. Mass Spectrom.* 238, 245 (2004).
44. P. M. Peacock and C. N. McEwen, *Anal. Chem.* 78, 3957 (2006).
45. S. M. Weidner and S. Trimpin, *Anal. Chem.* 80, 4349 (2008).
46. S. M. Weidner and S. Trimpin, *Anal. Chem.* 82, 4811 (2010).
47. T. Gruendling, S. Weidner, J. Falkenhagen, and C. Barner-Kowollik, *Polym. Chem.* 1, 599 (2010).
48. J. B. Fenn, M. Mann, C. K. Meng, S. F. Wong, and C. M. Whitehouse, *Mass Spectrom. Rev.* 9, 37 (1990).
49. M. Karas and F. Hillenkamp, *Anal. Chem.* 60, 2299 (1998).
50. K. Tanaka, H. Waki, Y. Ido, S. Akita, Y. Yoshida, and T. Yoshida, *Rapid Commun. Mass Spectrom.* 2, 151 (1988).
51. H. J. Griesser, P. Kingshott, S. L. McArthur, K. M. McLean, G. R. Kinsel, and R. B. Timmons, *Biomaterials* 25, 4861 (2004).
52. O. Vorm, P. Roepstorff, and M. Mann, *Anal. Chem.* 66, 3281 (1994).
53. D. C. Muddiman, A. I. Gusev, and D. M. Hercules, *Mass Spectrom. Rev.* 14, 383 (1995).
54. R. J. Cotter, B. D. Gardner, S. Litchenko, and R. D. English, *Anal. Chem.* 76, 1976 (2004).
55. D. C. Schriemer and L. Li, *Anal. Chem.* 68, 2721 (1996).

56. R. J. Cotter, B. D. Gardner, S. Litchenko, and R. D. English, *Anal. Chem.* 76, 1976 (2004).
57. D. C. Schriemer and L. Li, *Anal. Chem.* 68, 2721 (1996).
58. H. Pasch, R. Unvericht, and M. Resch, *Angew. Makromol. Chem.* 212, 191 (1993).
59. V. Schädler, J. Spickermann, J. Räder, and U. Wiesner, *Macromolecules* 29, 4865 (1996).
60. J. Spickermann, H. J. Räder, K. Müllen, B. Müller, M. Gerle, K. Fischer, and M. Schmidt, *Macromol. Rapid Commun.* 17, 885 (1996).
61. W. S. Wiley and I. H. McLaren, *Rev. Sci. Instrum.* 26, 1150 (1955).
62. R. S. Brown and J. J. Lennon, *Anal. Chem.* 67, 1998 (1995).
63. G. Goizueta, T. Chiba, and T. Inoue, *Polymer* 34, 253 (1993).
64. J. H. Butler, D. C. Joy, G. F. Bradley, and S. J. Krause, *Polymer* 36, 1781 (1995).
65. L. C. Sawyer and D. T. Grubb, *Polymer Microscopy,* Chapman & Hall, London (1996).
66. K. Osaki, M. Tamura, M. Kurata, and T. Kotaka, *J. Phys. Chem.* 69, 4183 (1965).

Chapter 6

Thermoplastics

Plastics are based on polymers and may include various other chemicals such as additives, stabilizers, colorants, processing aids, etc. Therefore, each product has to be optimized with respect to its processing and future application [1,2]. Thermoplastics—polymers that become liquid when heated and hard when cooled—are being used more and more in structural applications. The thermoplastics include basic polymers such as polyethylene (PE), polypropylene (PP), polystyrene (PS), and polyvinylchloride (PVC), and specialty resins such as acrylonitrile-butadiene-styrene (ABS), polymethylmethacrylate (PMMA), and high-density polyethylene (HDPE) as higher-order derivatives. Thermoplastics are particularly popular today for many applications because they are recyclable with minimum degradation upon exposure to heat.

The most vital element in the growth of instrumental techniques is found in their breadth of application. These methods provide a wealth of information that can be used in determining applications and thereby facilitate the continued growth in the use of thermoplastic materials. Further, instrumental methods play an important role in furthering our understanding of materials behavior and helping in the design of products that meet desired targets of performance and durability. Some representative data for a sampling of common thermoplastic materials are discussed in this chapter.

6.1. Polyethylene (PE)

Polyethylene is one of the most important materials in the plastics industry. Using controlled branching, polyethylenes can be developed with bimodal or

DOI: 10.5643/9781606502440/ch6

multimodal molecular-weight distributions (MWD)s. These materials offer both good physical properties and excellent processability [3,4]. Polyethylene can be found in several grades, and its consumption is constantly growing [5]. Common PEs include the following.

- *Low-density polyethylene* (LDPE) is a homopolymer produced in a high-pressure, noncatalyzed process. It contains both long and short branches. LDPE melts at low temperatures.
- *High-density polyethylene* (HDPE), consists of unbranched molecular chains. It has higher melting temperature and is also more fragile than LDPE.
- *Linear low-density polyethylene* produced by copolymerization of ethylene with other olefins such as 1-hexene. Together with the main chain, the large quantity of short branches produces properties that are intermediate between those of HDPE and LDPE.

6.1.1.   Infrared and Raman Spectra of Linear Polyethylene

Instrumental methods are commonly utilized by the polymer industry to determine an analytical strategy and interpretation. Some of the material characterization of polyethylene is given below.

The vibrational patterns of Raman and IR spectroscopy can be used of great advantage in the measurement of the structure of molecules. The more symmetric the molecule, the greater will be the differences between the Raman and IR spectra. The polyethylene molecule has a center of symmetry, so the Raman and IR spectra should exhibit entirely different vibrational modes. Figure 6.1 [6] shows the differences between the Raman and IR spectra for linear polyethylene. The C–C mode can be clearly observed in the Raman spectrum, but the CH_2 modes dominate the IR spectrum [6–8].

6.1.2.   Differential Scanning Calorimetry (DSC) of High-Density Polyethylene

Figure 6.2 shows a heating thermogram of pure HDPE [9]. The single peak seen is characteristic of semicrystalline polymers. The temperature corresponding to the peak represents the melting point (T_m) of the polymer. Table 6.1 lists the melting points of some commercial polymers and indicates that

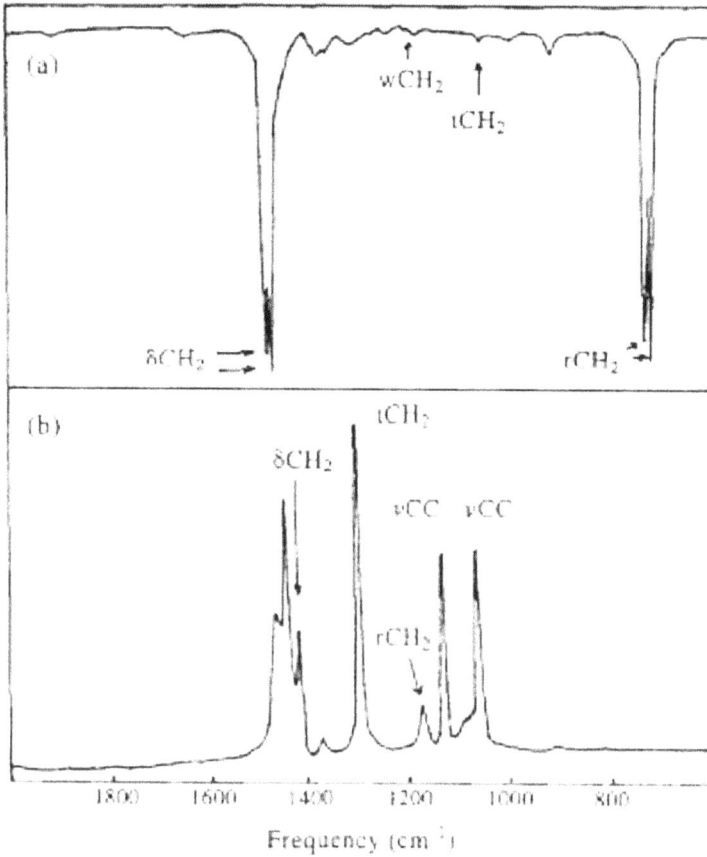

Figure 6.1. (a) Infrared and (b) Raman spectra of linear polyethylene. [Reprinted with permission from G. Xue, *Prog. Polym. Sci.* 22, 313 (1997). Copyright © 1997 Elsevier Science Ltd. All rights reserved.]

polypropylene has a high melting temperature, whereas that of LDPE is quite low [9].

6.1.3.   Fourier-Transform IR (FTIR) Spectra of LDPE

Figure 6.3 shows spectra obtained using three different techniques of sample preparation, in transmittance and attenuated total reflection (ATR) [10]. Line A, using transmission and a thin film, shows very intense bands, causing opaque spectral intervals. This effect is seen even with very thin films. Line B, using transmission and a potassium bromide (KBr) disk, indicates

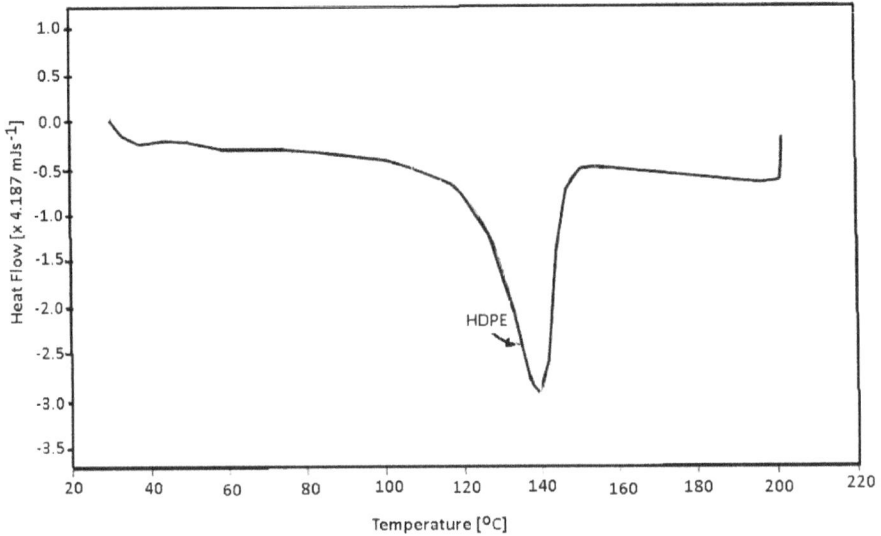

Figure 6.2. Typical DSC thermogram of HDPE. [Reprinted with permission from A. C.-Y. Wong and F. Lam, *Polym. Testing* 21, 691 (2002). Copyright © 2002 Elsevier Science Ltd. All rights reserved.]

a nonhomogeneous material, due to the softness of polyethylene and the mismatch of polarity between the two components. The KBr disk also acts as a diluent, decreasing the strong absorptions in the 3000 cm^{-1} region. As shown in line C, the ATR technique affords well-resolved and less intense bands similar to those obtained with the KBr disk. The advantage of ATR is the possibility of obtaining spectra directly from the sample, without any

Table 6-1 Melting Temperatures of Some Pure
Thermoplastics as Measured by DSC

Thermoplastic	Melting temperature (°C)
Polypropylene	169.4
LDPE	112.4
LLDPE	127.9
HDPE	138.4

Source: Reprinted with permission from A. C.-Y. Wong and F. Lam, *Polym. Testing* 21, 691 (2002). Copyright © 2002 Elsevier Science Ltd. All rights reserved.

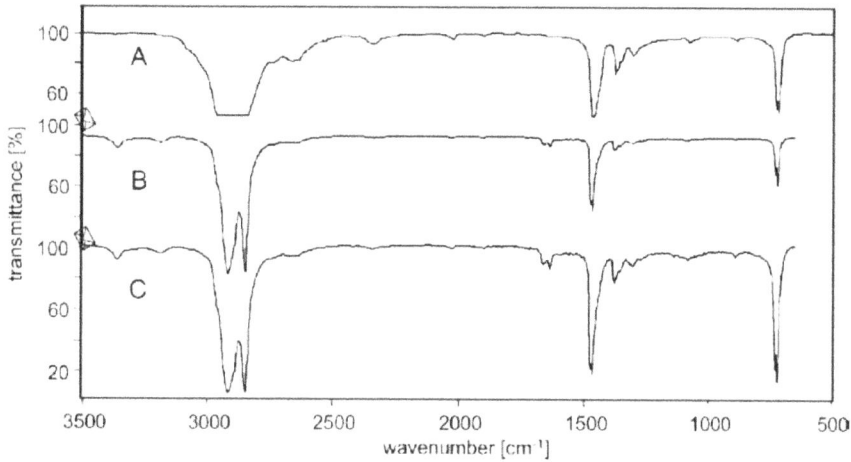

Figure 6.3. LDPE spectra acquired using different FTIR techniques: A, transmission—film; B, transmission—KBr disk; C, ATR obtained with ZnSe and 45° incident beam. [Reprinted with permission from J. V. Gulmine, P. R. Janissek, H. M. Heise, and L. Akcelrud, *Polym. Testing* 21, 557 (2002). Copyright ©2002 Elsevier Science Ltd. All rights reserved.]

sample preparation. The same fundamental vibrations are detected, with only variations in intensity.

6.2. Polypropylene (PP)

Polypropylene isprobably the most typical example of a major commercial plastic. The practical use of PP was achieved with the development of adequate stabilizer systems. PP needs protection in every stage of its life cycle. The requirements for a stabilizer for PP are similar to those of PE but the stabilizer levels for PE are usually lower because PE is inherently less sensitive to oxidative attack than PP. PP is found in three isomeric forms, of which the isotactic form is the only one of commercial significance, the others being syndiotactic and atactic. By studying spectral changes with correlation analysis, one can often elucidate complex polymeric structural and morphological information on semicrystalline PP polymers.

6.2.1. FTIR Spectrum of Polypropylene

Incontrasttothe spectrum in Figure 6.1, the IR spectrum of linear polyethylene, the FTIR spectrum of polypropylene, shown in Figure 6.4 [11], indicates a

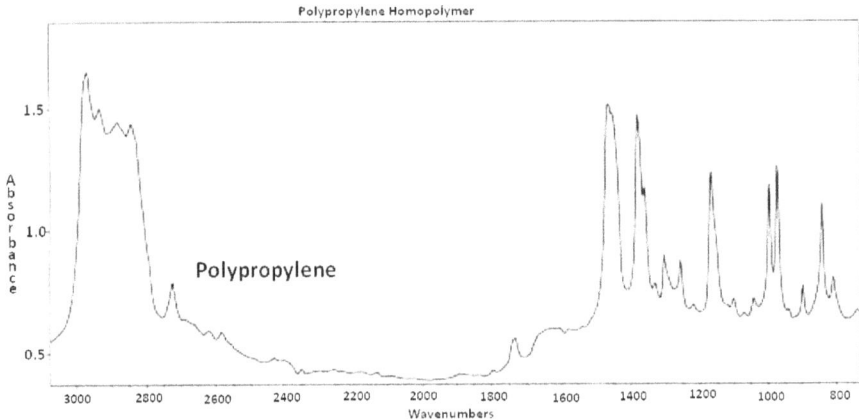

Figure 6.4.   FTIR spectrum of polypropylene homopolymer. [Reprinted with permission from A. Riga, R. Collins, and G. Mlachak, *Thermochim. Acta* 324, 135 (1998). Copyright © 1998 Elsevier Science B.V. All rights reserved.]

shoulder at 2875 cm^{-1}, and the asymmetric and symmetric in-plane C–H (–CH$_3$) at 1455 and 1358 (shoulder) confirm polypropylene.

6.2.2.   FTIR Spectra of Degradation of Polypropylene

The degradation of polypropylene shows random chain scission. It results in oligomers of 2,3-dimethyl-1-hexene (C$_8$H$_{16}$), 2-ethyl-1-butene (C$_6$H$_{12}$), 2-methyl-1-pentene (C$_6$H$_{12}$), 2-ethyl-1-hexene (C$_8$H$_{16}$), and 2- methyl-1-butene (C$_5$H$_{10}$) [12] according to peaks at 3084, 2964, 1651, 1460, 1377, and 889 cm^{-1} and oxidized species (1730 cm^{-1} signals the C–O stretching vibration) and also CO, CO$_2$, and H$_2$O, as shown in Figure 6.5 [13]. Absorbance of the selected peaks in Figure 6.5b is shown at 889 cm^{-1} [C–H deformation vibration (alkene C–H vibration)], 1730 cm^{-1} (C–O stretching vibration), and 2964 cm^{-1} [C–H stretching vibration (–CH$_3$)] as a function of time. The length of the heating stage depends on the properties of the sample.

6.3.   Polystyrene (PS)

Polystyrene belongs to the group of standard thermoplastics. It has special properties which mean it can be used in an extremely wide range of applications. Polystyrene materials are hard, transparent materials with a high gloss. Above its softening point, clear PS occurs as a melt and can be processed by processing techniques such as injection molding or extrusion. Below 100°C,

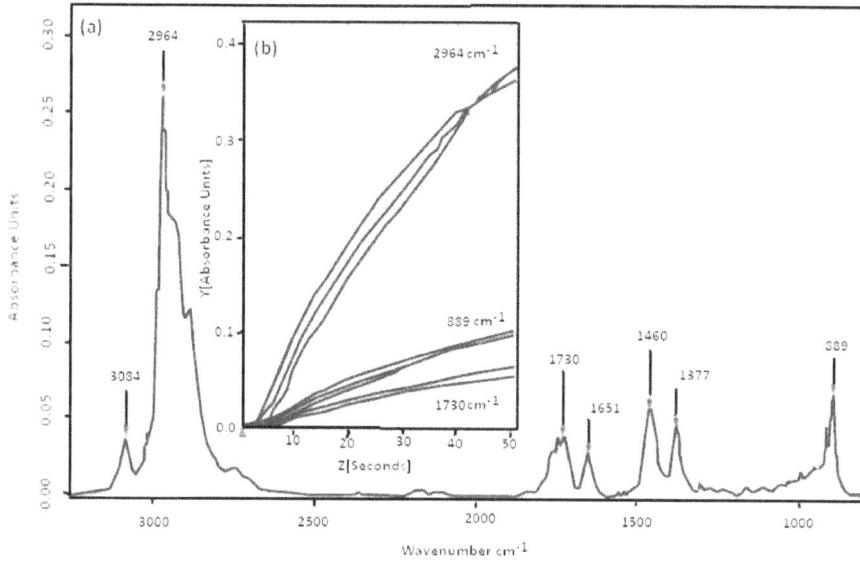

Figure 6.5. Traces of selected peaks of polypropylene decomposition spectra (a) at 889 cm⁻¹, (b) 1730 cm⁻¹, and (c) 2964 cm⁻¹ as a function of time. [Reprinted with permission from B. Bodzay, B. B. Marosfoi, T. Igricz, K. Bocz, and G. Marosi, *J. Anal. Appl. Pyrolysis* 85, 313 (2009). Copyright © 2009, Elsevier. All rights reserved.]

polystyrene materials solidify to give a glasslike material with adequate mechanical strength, good dielectric properties, and resistance toward a large number of chemicals for many types of applications.

Polystyrene is one of the most important commercial polymers today [14]. Its popularity stems from the fact that it possesses many useful properties, such as good processability, rigidity, transparency, and low water absorbability, and it can be produced at low cost [15]. However, it also has some disadvantages, which are detailed below.

6.3.1.   Molecular-Weight Distribution of Polystyrene

Various approaches have been used to overcome the disadvantages of polystyrene, including the addition of comonomers, chain transfer agents, or cross-linking agents before polymerization [16]. Sometimes, saturated monomers are used for reinforcement [17].

In gel permeation chromatography, pretreatment of the polymers is necessary to obtain narrow fractions. To obtain Figure 6.6, separation of the

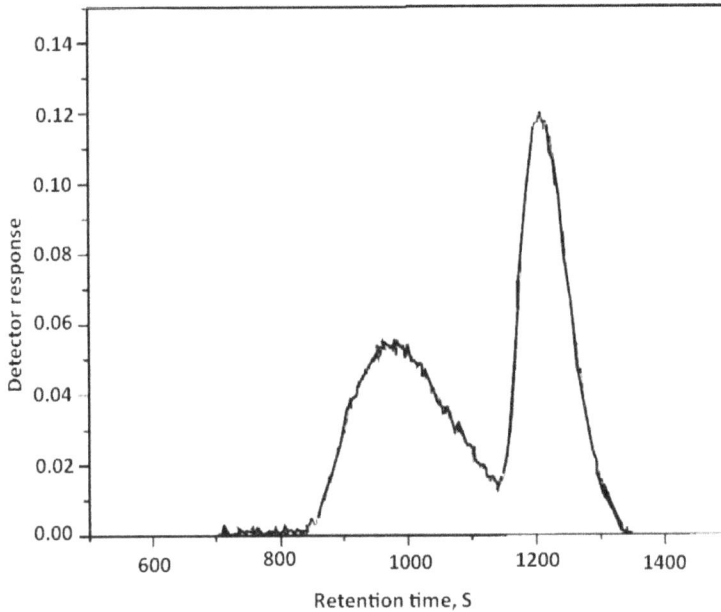

Figure 6.6. Molecular-weight distribution of a mixture of polystyrene and polyvinyl acetate. [Reprinted with permission from V. Karmore and G. Madras, *Ind. Eng. Chem. Res.* 40, 1307 (2001). Copyright © 2001, American Chemical Society.]

polymers was carried out using gel permeation chromatography by dissolving the polymer in a solvent and then reprecipitating, followed by removal of the solvent [18]. Figure 6.7 is a calibration curve obtained by injecting polystyrene standard [18]. The curve is used to convert retention time to molecular weight. The peaks in Figure 6.6 for both polystyrene and polyvinyl acetate from the mixture are distinct.

6.3.2. Thermogravimetric Analysis (TGA) of Polystyrene

Polystyrene degradation proceeds via a chain scission pathway [20]. The polystyrene radical then undergoes scission of the polystyrene C–C bond. TGA data for the degradation of polystyrene in Figure 6.8 indicate that the onset of degradation starts experimentally at about 341°C and suggest that breakage of the C–C bonds for polystyrene degradation (273 kJ/ mol) starts occurring at about this temperature [19].

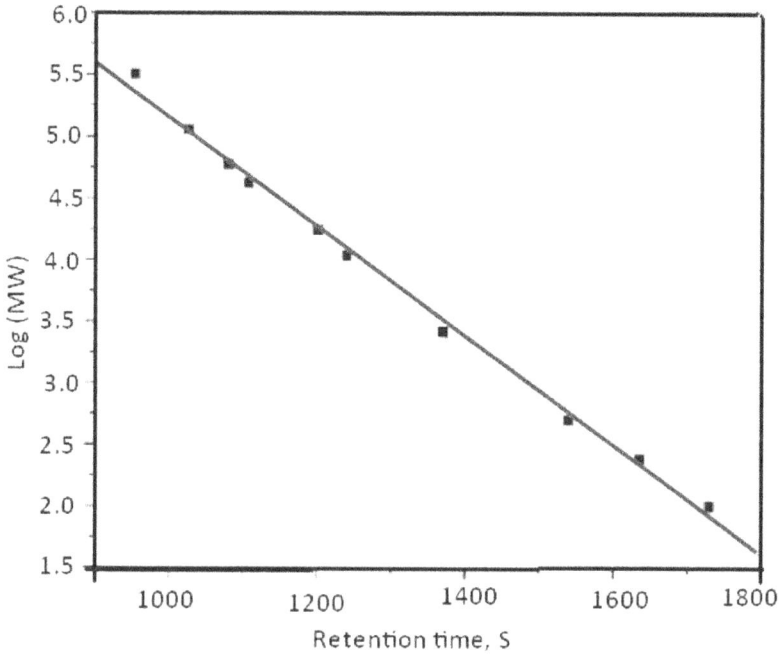

Figure 6.7.   Calibration curve based on polystyrene standard. [Reprinted with permission from V. Karmore and G. Madras, *Ind. Eng. Chem. Res.* 40, 1307 (2001). Copyright © 2001, American Chemical Society.]

6.4.   Polyethylene Terephthalate (PET)

Polyethylene terephthalate is the most common and important commercial polyester. Several modifications of it have now been systematically investigated by different instrumental methods of analysis. The influence of chemical structure on physical properties is used structurally to modify polyesters containing different acid and glycol monomers. PET is thermally and mechanically treated to determine the effects of such treatments on the transition. Some instrumental analysis results are given below.

6.4.1.   Infrared and Raman Spectra of Polyethylene Terephthalate

Figure 6.9 shows the IR and Raman spectra of polyethylene terephthalate [8]. The C=C stretching modes of the aromatic ring clearly dominate the

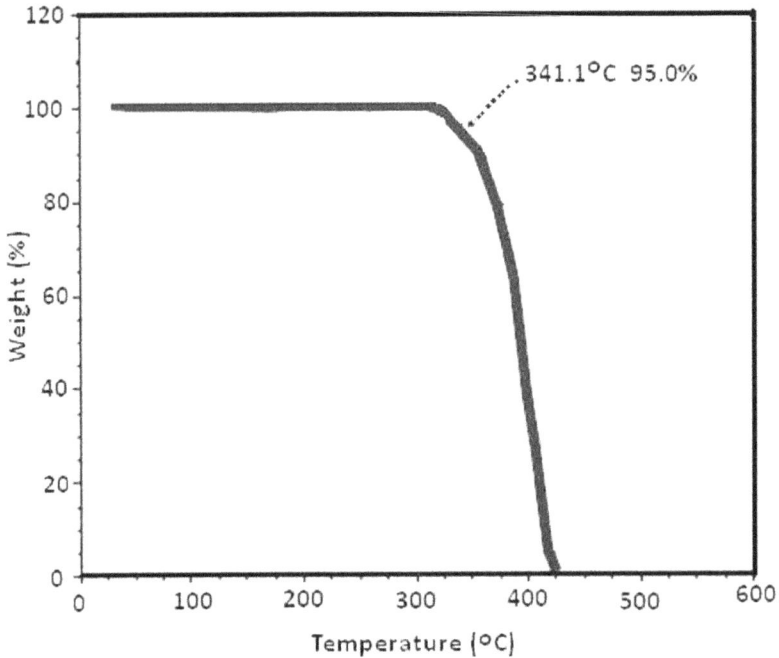

Figure 6.8. TGA curve of pure polystyrene. [Reprinted with permission from M. W. Beach, N. G. Rondan, R. D. Froese, B. B. Gerhart, J. D. Green, B. G. Stobby, A. G. Shmakov, V. M. Shvartsberg, and O. P. Korobeinichev, *Polym. Degrad. Stabil.* 93, 1664 (2008). Copyright © 2008 Elsevier Ltd. All rights reserved.]

Raman spectrum, and in the IR spectrum the C–O stretching modes are the strongest. The Raman spectrum has a fluorescence background that makes the dynamic range less than is desirable. The Raman lines are rather weak when superimposed on the broad background.

6.4.2.   NMR Spectrum of Polyethylene Terephthalate

The ^{1}H-NMR spectra of pure PET and a deuterated chloroform-soluble fraction of thermally treated glycol-modified polyethylene terephthalate/liquid crystalline polymer (PETG/LCP) (70:30 wt/wt) blend are shown in Figure 6.10 [23]. There is no difference in the peaks, indicating that no transesterification occurs. The crystallization rate of PET in LCP is significantly higher than that of pure PET. Therefore the LCP polymer appears to act as a nucleating agent for PET [21,22].

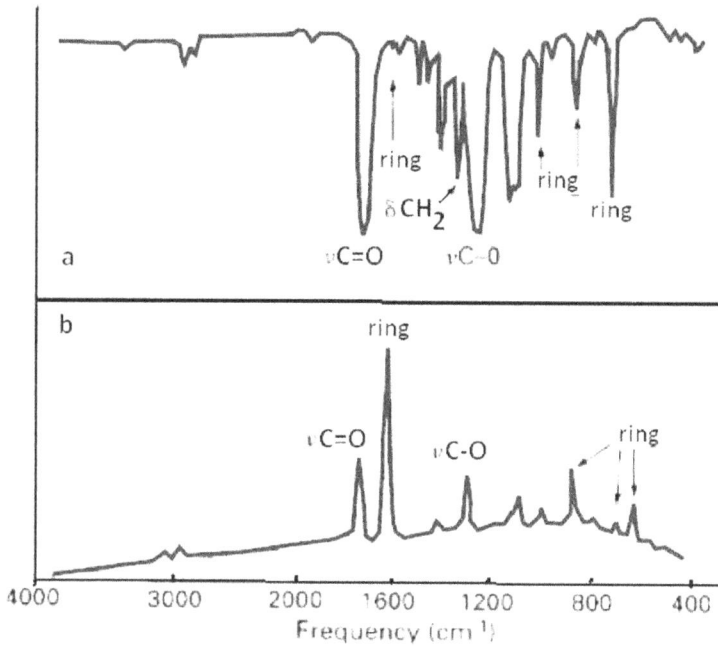

Figure 6.9. (a) Infrared and (b) Raman spectra of polyethylene terephthalate. [Reprinted with permission from G. Xue, *Prog. Polym. Sci.* 19, 317 (1994). Copyright © 1997 Elsevier Science Ltd. All rights reserved.]

6.4.3. Differential Scanning Calorimetry of Polyethylene Terephthalate

Figure 6.11 [24] shows that the fusion temperature (T_f) and crystallization temperature (T_c) for PET are at 231 and 201°C, respectively, with a glass transition temperature (T_g) of 80°C. The average molecular weight and polydispersity of PET are comparable, as shown by the presence of a single peak at the melting point [24].

6.5. Polyvinylchloride (PVC)

Polyvinylchloride (PVC) is a linear, thermoplastic, substantially amorphous polymer, with a huge commercial interest, due to the accessibility to basic raw materials and to its properties [25]. When PVC is plasticized, it presents some interesting properties which make it widely accepted for use in

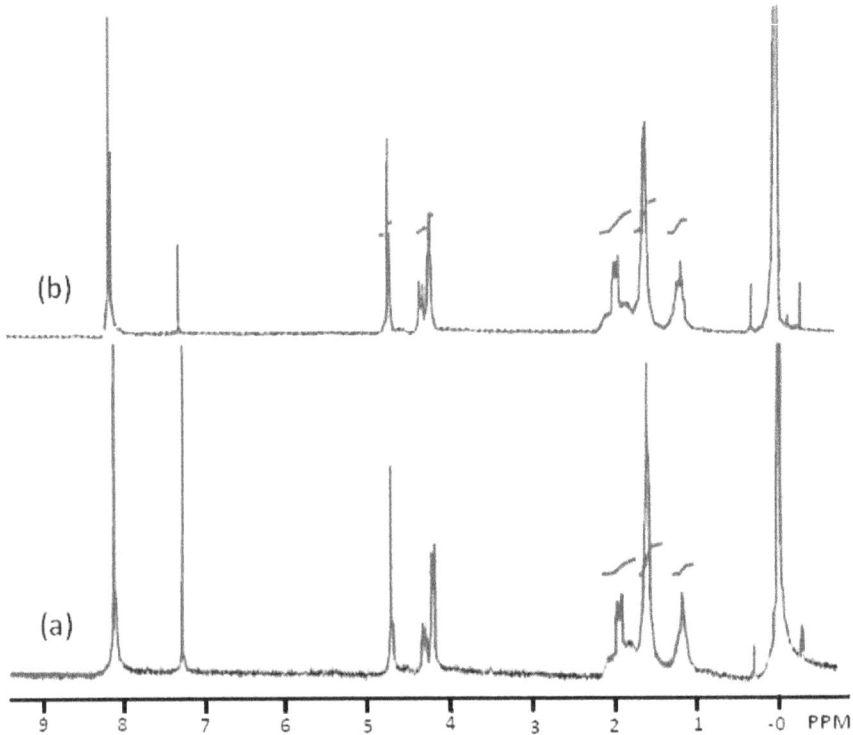

Figure 6.10. ¹H-NMR spectra of pure (a) PET and (b) deuterated chloroform (CDCl₃)-soluble fraction of PETG/LCP (70:30 wt/wt) thermally treated blend. [Reprinted with permission from S.-H. Hwanga, K.-S. Jeong, and J.-C. Jung, *Eur. Polym. J.* 35, 1439 (1999). Copyright © 1999 Elsevier Science Ltd. All rights reserved.]

flexible medical products (dialysis, blood, urine and secretion bags, blood tubing for hemodialysis, endotracheal tubes, intravenous solution dispersion sets, catheters, contact lenses, gloves, as well as for drug product storage and packaging) [26]. In addition, many other PVC medical devices have passed critical toxicological, biological, and physiological tests [27].

PVC is regarded as second behind polyethylene in terms of worldwide polymer consumption, but recycling of postconsumer PVC is not as widespread as for polyethylene because of its long-term applications. PVC can be used in a variety of applications, including bottles, drainage pipes, and drainage pipe fittings, with good appearance and properties. PVC is a proton-donating polymer that interacts with oxirane rings of epoxidized oils and hydrocarbons, resulting in the plasticization of the resin. Thus epoxidized oils have been used as partial replacements for plasticizers, which also act as thermal stabilizers for PVC [28]. These plastics are identified by

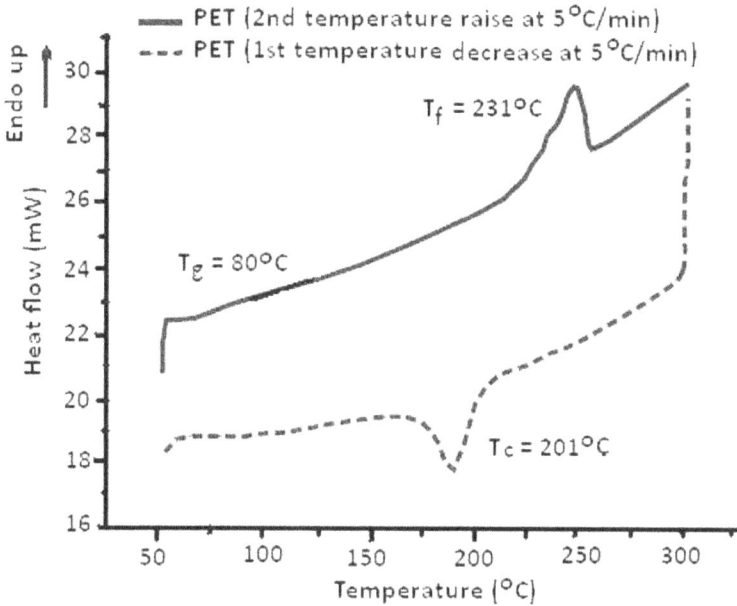

Figure 6.11. DSC of PET. [Reprinted with permission from G. Colomines, A. van der Lee, J.-J. Robin, and B. Boutevin, *Eur Polym. J.* 44, 2874 (2008). Copyright © 2008 Elsevier Ltd. All rights reserved.]

instrumental methods of analysis, and some instrumental results for PVC are presented as examples of our knowledge about thermoplastic materials.

6.5.1.   FTIR Spectrum of Polyvinylchloride

An FTIR spectrum of pure PVC is shown in Figure 6.12 [29]. A C–H stretching mode can be observed at 2911 cm^{-1}, a CH$_2$ deformation mode at 1333 cm^{-1}, a C–H rocking mode at 1254 cm^{-1}, a *trans*-CH wagging mode at 959 cm^{-1}, a *cis*-CH wagging mode at 616 cm^{-1}, and C–Cl stretching at 844 cm^{-1} [30–32].

6.5.2.   FTIR Spectrum of PVC and Plasticizer

For PVC with a plasticizer, the FTIR spectrum shows, as expected, some characteristic bands of PVC and some of the plasticizer, as can be seen in Figure 6.13 [33]. For example, the bands at 715–556 and 1444–1414 cm^{-1} are attributed, respectively, to the C–Cl and C–H bonds of PVC, while the bands at 1477–1444 and 1803–1655 cm^{-1} are attributed, respectively, to the methyl (–CH$_3$) and carbonyl groups of the plasticizer.

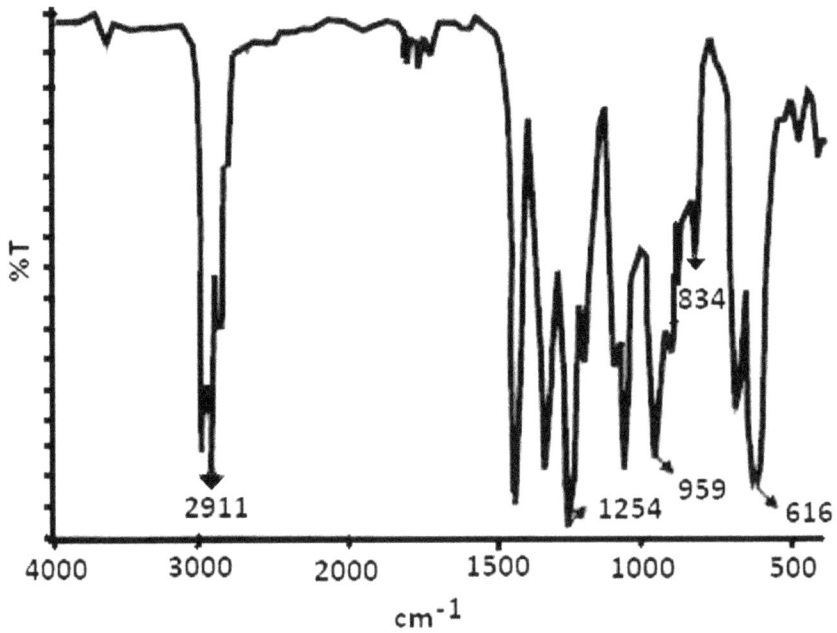

Figure 6.12.   FTIR spectrum of pure PVC. [Reprinted with permission from S. Ramesh, K. H. Leen, K. Kumutha, and A. K. Arof, *Spectrochim. Acta A* 66, 1237 (2007). Copyright © 2007 Elsevier B.V. All rights reserved.]

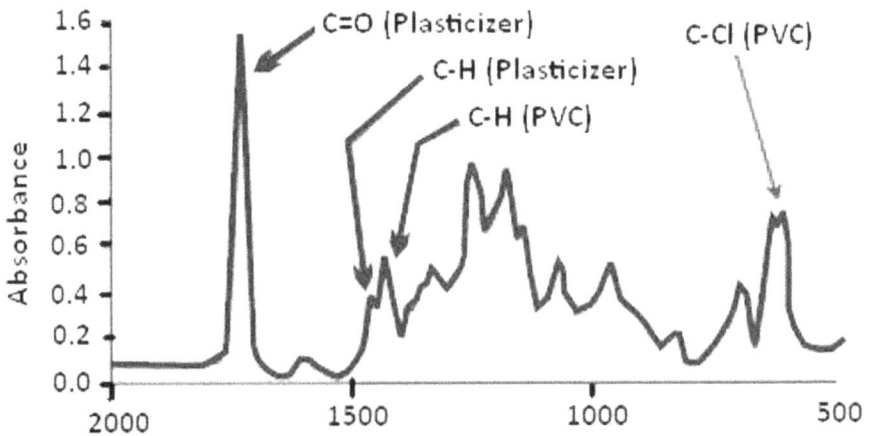

Figure 6.13.   Characteristic absorption bands of PVC with a plasticizer. [Reprinted with permission from A. Marcilla, S. Garcia, and J. C. Garcia-Quesada, *Polym. Testing* 27, 221 (2008). Copyright © Elsevier Ltd. All rights reserved.]

6.6.   Polymethylmethacrylate (PMMA)

Polymethylmethacrylate is widely used in many industrial fields due to its physical and mechanical properties, such as high optical transmission in the infrared and visible ranges, easy thermoforming, and good stability with time. Therefore, the improvement of its behavior toward thermal degradation is a challenge to be solved in order to increase the fire-retardation properties and the commercial value of this material. Instrumental methods of analysis have entered an era in which great emphasis is now being placed on the applications of existing polymers. With the help of instrumental methods, the art of tailoring polymers has emerged to obtain some synergistic effect. In light of this importance, some of the work on fundamental relationships using instrumental methods of analysis is presented.

6.6.1.   Molecular-Weight Distribution of PMMA

The mass MWD in Figure 6.14 indicates a number-average molecular weight of 300 and a weight-average molecular weight of 11,900 for PMMA [34]. The

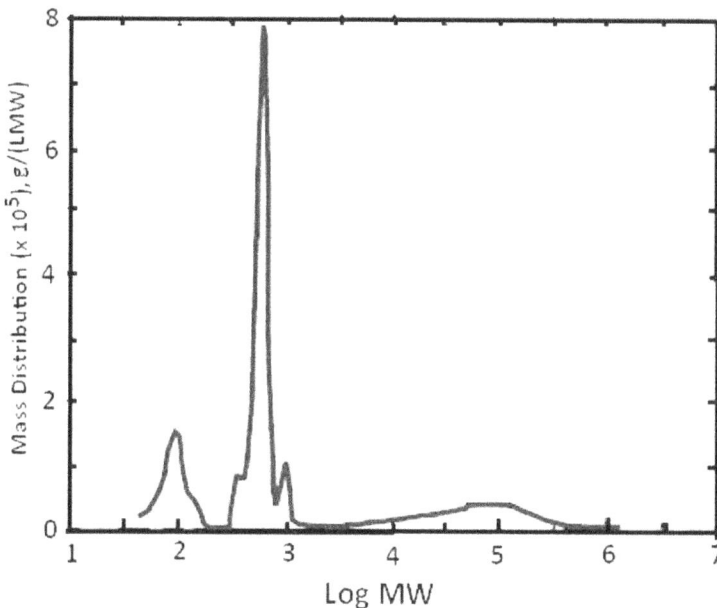

Figure 6.14.   Molecular-weight distribution of PMMA (MN = 300, MW = 11,900). [Reprinted with permission from G. Madras, J. M. Smith, and B. J. McCoy, *Ind. Eng. Chem. Res.* 35, 1795 (1996). Copyright © 1996 American Chemical Society. All rights reserved.]

graph shows many peaks in the lower MW range of the raw PMMA that may interfere with the product(s) of degradation.

6.6.2.   Infrared and Raman Spectra of PMMA

The IR and Raman spectra of PMMA are compared in Figure 6.15 [6]. The C=C and C–O bands dominate the IR spectrum, and the C–C modes dominate the Raman spectrum. The Raman spectrum is particular rich in the lower frequency range, while there is little absorbance in this region of the IR spectrum. A single Raman instrument can be used to scan the region from 10 to 4000 cm^{-1}. FTIR spectroscopy, on the other hand, requires changing the beam splitter and detector to reach the far-IR region of the spectrum.

6.6.3.   TGA and DSC of PMMA

The decomposition of PMMA proceeds in three general stages, as shown by the TGA/DSC output (Figure 6.16) [35]. From the DSC data, the first stage

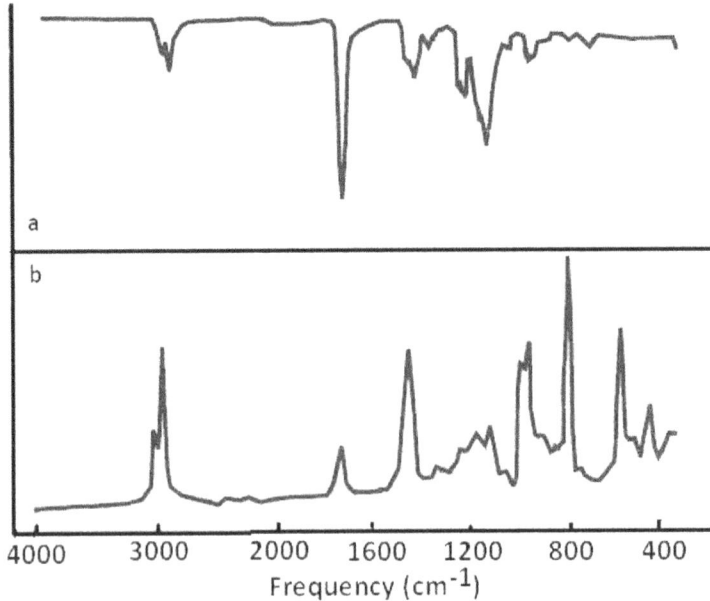

Figure 6.15.   (a) Infrared and (b) Raman spectra of PMMA. [Reprinted with permission from G. Xue, *Prog. Polym. Sci.* 19, 317–388 (1994). Copyright © 1994 Elsevier Ltd. All rights reserved.]

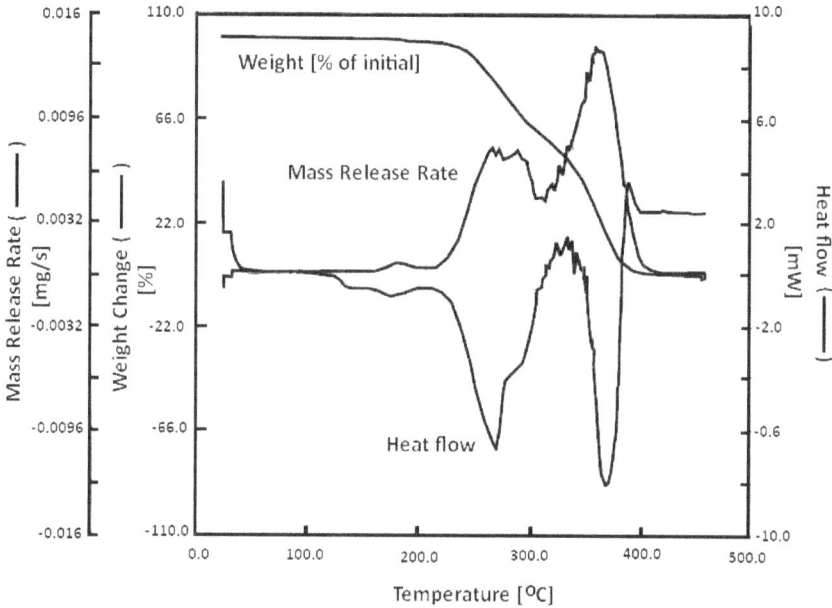

Figure 6.16. Typical TGA/DSC data for PMMA. [Reprinted with permission from P. R. Westmoreland, T. Inguilizian, and K. Rotem, *Thermochim. Acta* 367–368, 401 (2001). Copyright © 2001 Elsevier Science B.V. All rights reserved.]

begins at about 120°C, although the small detectable mass loss (2%) and its corresponding thermal effect occur only from 150 to 200°C in this stage. The second stage destroys about 40% of the sample, most rapidly at 270°C, with an apparent change of mechanism at about 290°C. Finally, the sample is completely decomposed by 410°C, reaching a maximum in rate of mass loss and heat uptake at 370°C.

6.7. Polyvinyl Acetate (PVAc)

Polyvinyl acetate is one of the most important synthetic polymers in wide use for industrial applications, such as adhesives and coatings for papers and textiles. PVAc has a low glass transition temperature (30°C) and good adhesion properties, and is applied as a plasticizer and component of coatings, paints, and glues.

PVAc is a cost-efficient, high-tonnage bulk commodity polymer [36]. Because of its low mechanical strength, PVAc cannot be used alone. Therefore, it should be used in combination with other polymers in the form of a copolymer. As a well-known example, copolymerization of PVAc with polyethylene forms ethylene vinyl acetate (EVA). Because of its mechanical properties, chemical resistance, flexibility, and processability, EVA is used extensively in many engineering and industrial fields [37,38].

When it is heated, PVAc liberates acetic acid, and conjugated double bonds are formed in the residual polymer. The degradation proceeds by a nonradical chain mechanism initiated at chain ends and propagated from unit to unit along the chain. The loss of weight is measured at constant temperatures between 230 and 300°C. Acetic acid is the most abundant volatile product formed (90–95%).

6.7.1.   FTIR Spectra of PVAc

From Figure 6.17, it can be observed that the CH_2 wagging of pure PVAc is centered at 1377 cm^{-1} [39]. The peak at 1736 cm^{-1} is attributed to C=O

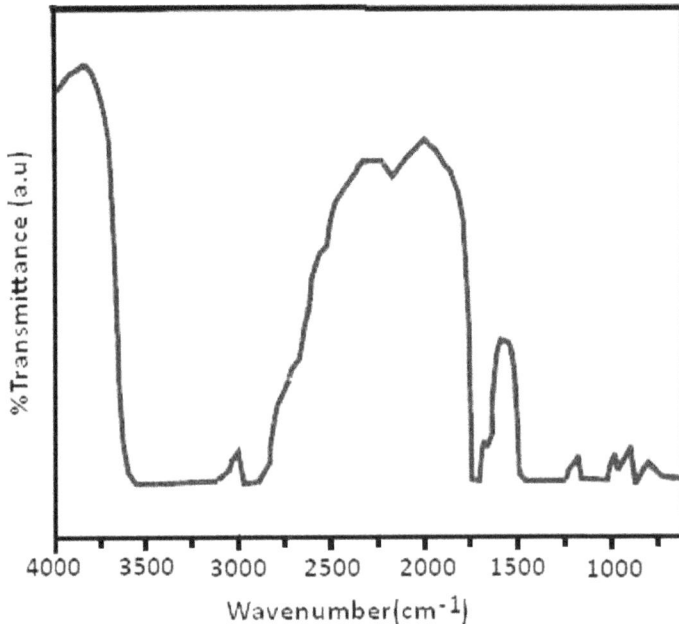

Figure 6.17. FTIR spectrum of pure PVA. [Reprinted with permission from M. Hema, S. Selvasekarapandian, D. Arunkumar, A. Sakunthala, and H. Nithya, *J. Non-Crystalline Solids* 355, 84 (2009). Copyright © 2009 Elsevier. All rights reserved.]

stretching of pure PVAc. The peak at 850 cm^{-1} corresponds to skeletal C–H rocking of pure PVAc [40].

6.7.2.   FTIR Spectra of PVAc and Its First-Stage Degradation

Figures 6.18a and 6.18b show PVAc spectra before and after the first stage of degradation. For comparison, Figures 6.18c and 6.18d show polyvinylchloride (PVC) spectra before and after the first stage of degradation. The peaks at 2964.6, 2866.4, 1434.1, and 1371.8 cm^{-1} in Figure 6.18a indicate different modes of vibration of CH$_2$ and CH$_3$. The peaks at 1737.8 and 1243.5 cm^{-1} correspond to C=O and C–O, respectively, suggesting the acetate structure of PVAc [41].

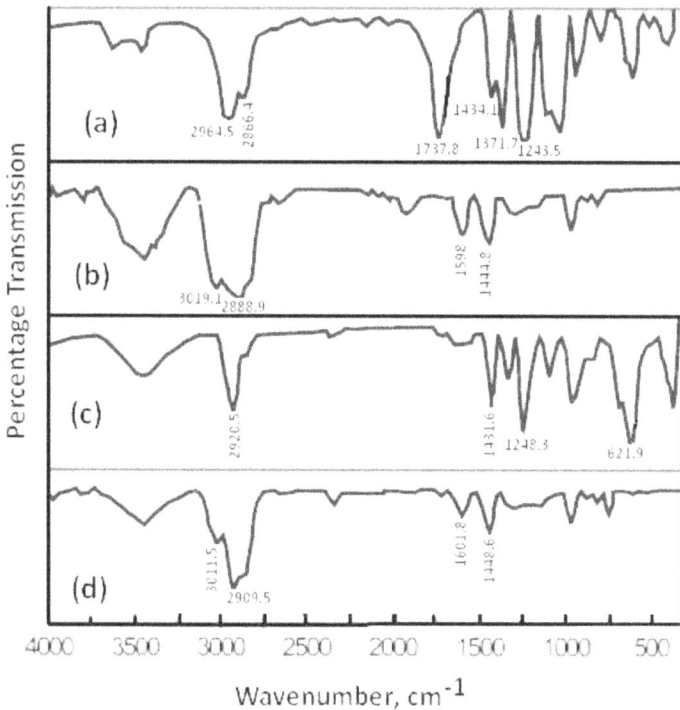

Figure 6.18. FTIR spectra of (a) PVAc, (b) PVAc after first-stage degradation, (c) PVC, and (d) PVC after first-stage degradation. [Reprinted with permission from G. Sivalingam, R. Karthik, and G. Madras, *Ind. Eng. Chem. Res.* 42, 3647 (2003). Copyright © 2003, American Chemical Society.]

Figure 6.18b shows an absence of C=O and C–O peaks, indicating the absence of acetate groups, and the peak at 3019.1 cm^{-1} indicates the formation of an olefinic C–H bond. Figure 6.18c shows the presence of a C–Cl group at 621.9 cm^{-1} in addition to the usual functional groups of CH$_2$ vibration modes. Figure 6.18d shows the absence of the C–Cl group and also the appearance of the olefinic bond and functional groups similar to those of Figure 6.18b. Thus, one can conclude that the skeleton left after the first degradation of both polymers is similar and that the olefinic breakage is relatively unaffected.

6.8.   Nylon

Nylon is the common name for synthetic polyamides. Among the commercially available thermoplastics, nylon is an important semicrystalline thermoplastic polymer having good mechanical, thermal, and chemical properties. In the automotive industry, nylon is widely used in high-performance applications such as in radiator end tanks, engine covers, intake manifolds, dipstick caps, thrust washers, liners for sheathed cables, etc. [42]. Extended exposure of nylon-based products to physical factors such as thermal oxidation, radiation, etc., and chemical factors is unavoidable and can lead to degradation. Instrumental methods are used to focus on such effects, and some results are given below.

6.8.1.   FTIR Spectra of Nylon 6

The spectrum of nylon 6 (Figure 6.19) shows the characteristic double peak corresponding to carbon–carbon bonds in the organic skeleton (2931 and 2862 cm^{-1}) [43]. The two significant peaks at 1635 and 1543 cm^{-1} are attributed to the secondary amide group, which is strongly represented in nylon 6. The peak at 980 cm^{-1}, although much weaker than the others, cannot be neglected, and is assigned to the carboxylic acid group.

6.8.2.   FT Raman Spectra of Nylon 4 and Nylon 11

Good-quality FT Raman spectra of nylon 4 and nylon 11 are shown in Figure 6.20 [6]. Assessed purely on a fingerprint basis, the spectra are very specific, and there is no difficulty identifying any member of the series. The ratio of the intensity of the methylene bending mode at 1440 cm^{-1} and the amide I mode at 1640 cm^{-1} is plotted against the number of methylene groups per

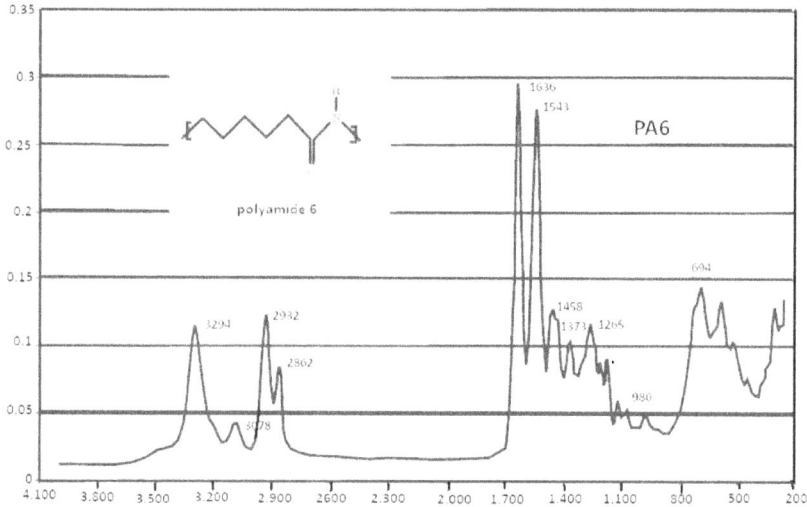

Figure 6.19. Spectra of nylon 6. [Reprinted with permission from K. Lamnawar, A. Baudoin, and A. Maazouz, *Eur. Polym. J.* 46, 1604 (2010). Copyright © 2010 Elsevier. All rights reserved.]

repeat unit and, excluding the CH_2 group adjacent to the carbonyl group, a straight line is obtained.

6.9. Polycarbonate (PC)

Polycarbonate resin has excellent mechanical strength, particularly impact strength, good electrical properties, and transparency. It is widely utilized in a variety of fields, for office machinery, electrical and electronic machinery, automobiles, architectural applications, and others. Many applications require that the polycarbonate be flame retardant and combine ease of processing with good optical properties.

Polycarbonate, especially aromatic PC, is a widely used thermoplastic resin due to its outstanding properties, including good hardness, stiffness, impact strength, transparency, dimensional stability, and thermal stability. The oxygen index of typical virgin bisphenol-A polycarbonate is about 25. However, more stringent flame-retardant performance is often required [44,45]. The use of instrumental methods of analysis is highly beneficial in this case, because the properties of the thermoplastic material can be improved. Some of the instrumental methods are given below in conjunction with the characterization of these thermoplastics.

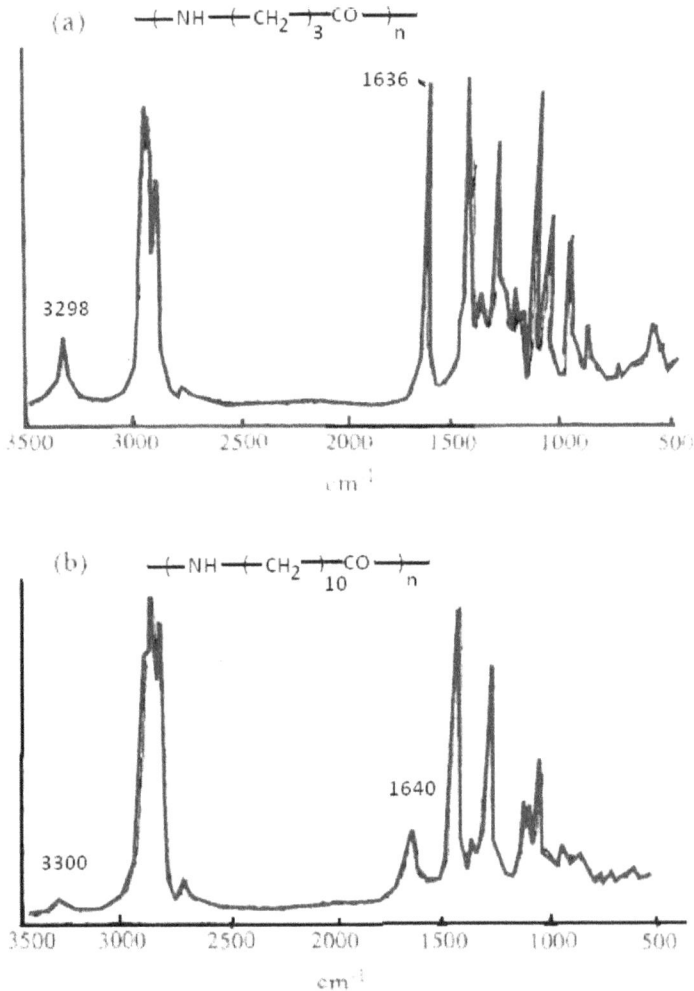

Figure 6.20. Fourier-transform Raman spectra of (a) nylon 4 and (b) nylon 11. [Reprinted with permission from G. Xue, *Prog. Polym. Sci.* 22, 313 (1997). Copyright © Elsevier Science Ltd., All rights reserved.]

6.9.1.   FTIR Spectrum of Polycarbonate

The FTIR spectrum of polycarbonate is shown in Figure 6.21 [43]; the assignments are listed in Table 6.2. Figure 6.21 shows expanded IR spectra between 500 and 4500 cm^{-1}. The peak at 1380 cm^{-1} corresponds to asymmetric and symmetric C–H(–CH$_3$), the band at 1779 cm^{-1} is the C=O stretching frequency of the carbonate linkage, and the 1200 cm^{-1} band corresponds to isopropylidene vibrations of polycarbonate. The peaks at 1593, 1506, and 1454 cm^{-1} (shoulder) are the C=Cs of the aromatic ring.

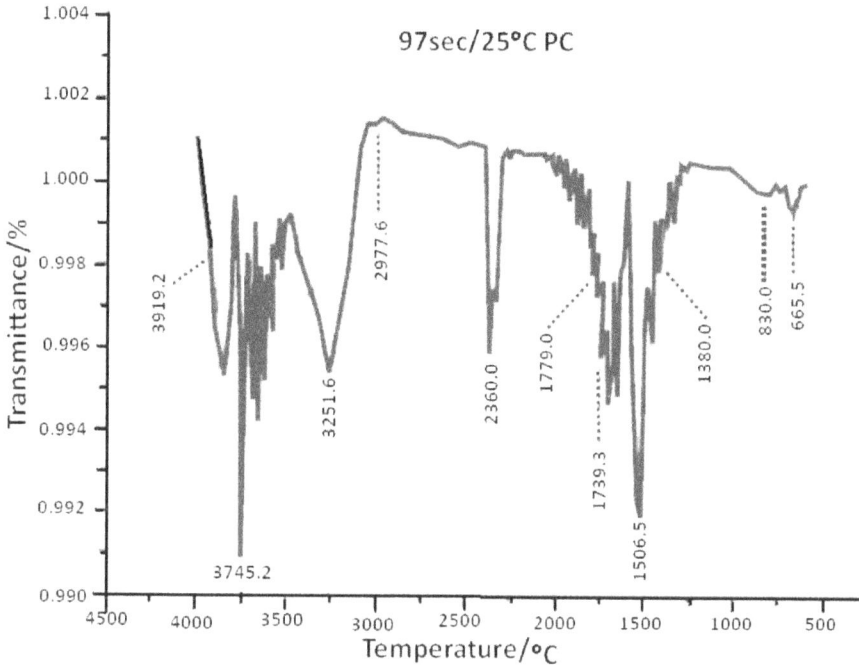

Figure 6.21.   FTIR spectrum of polycarbonate at 25°C. [Reprinted with permission from X. Haung, X. Ouyang, F. Ning, and J. Wang, *Polym. Degradation Stability* 91, 606 (2006). Copyright © 2006 Elsevier. All rights reserved.]

6.9.2.   Thermogravimetric Analysis of Polycarbonate

As shown in Figure 6.22 [46], the TGA of polycarbonate indicates that the melting temperature peaks at 494°C and the melting range starts around 346.5–542°C. At 423.8°C, a phase change occurs within the melting range. The percent residues remaining at 500 and 700°C are 40.1% and 21.5%, respectively.

6.10.   Infrared Bands for Identification of Thermoplastic Materials

The majority of the polymers used in commodity plastics have covalent bonding. In general, polymers belong to a very wide series of same and/or different monomers and also with different substrates. Table 6.2 illustrates some of the thermoplastic materials and their main IR bands for identification. This table is a good reference which provides a way to normalize spectral intensities.

Table 6-2 Thermoplastic Materials and Main IR Bands for Identification

Thermoplastic	Description	Main IR bands ν (cm^{-1})	Refs.
Low-density polyethylene (LDPE)	Asymmetric and symmetric –CH$_2$– Asymmetric and symmetric in-plane C–H Rocking (–CH$_2$–)	2915, 2845 1470, 1377 717	6,10
Polypropylene (PP)	Asymmetric and symmetric –CH$_3$ Asymmetric and symmetric –CH$_2$ Asymmetric and symmetric in-plane C–H(–C–H$_2$) Asymmetric and symmetric in-plane C–H(–C–H$_3$) Rocking –CH$_3$ and C–C Rocking and wagging –CH$_2$ C–C polymer backbone Rocking –CH$_2$–	2950, 2867 2917, 2838, 2875 (shoulder) 1455, 1358 (shoulder) 1436 (shoulder), 1375 1166, 972 997 1153 (shoulder), 808 898, 840	11,12
Polyethylene terephthalate (PET)	Asymmetric and symmetric –CH$_2$– C=O (ester) C=C (aromatic ring) Asymmetric and symmetric in-plane C–H C–C(O)–O –O–C– In-plane C–H aromatic ring Rocking C–H(–CH$_2$–) Wagging C–H aromatic ring	2965, 2923, 2908, 2853 1713 1615 (shoulder), 1580, 1505, 1454 1470, 1410, 1372, 1340 1240 1120 (shoulder), 1098 870 847 720	
Polycarbonate (PC)	=C–H aromatic ring Asymmetric and symmetric –CH$_3$ Asymmetric and symmetric –CH$_2$ C=O ester	3057, 3040 2950, 2872 2929, 2853 1770, 1593, 1506, 1454 (shoulder)	45

Polycarbonate (PC) (*continued*)	C=C aromatic ring	1464, 1364
	Asymmetric and symmetric in-plane C–H(–CH$_2$–)	1235, 1186, 1163
	O–C(O)–O	1409, 1380
	Asymmetric and symmetric C–H(–CH$_3$)	1102
	In-plane C–H aromatic ring	1013
	–O–C(O)–O out-of-plane C–H aromatic ring	886
	Wagging C–H aromatic ring	831
Polystyrene (PS)	=C–H aromatic ring	3081, 3059, 3024, 3001
	Asymmetric and symmetric –CH$_2$–	2919, 2847
	Overtone mono-substituted aromatic ring	1940, 1868, 1800, 1743
	C=C aromatic ring	1601, 1583, 1492, 1451
	Asymmetric and symmetric in-plane –C–H(–CH$_2$–)	1372 and 1311
	In-plane =C–H	1180, 1154, 1068, 1027
	Out-of-plane =C–H	905, 840, 749
	Out-of-plane aromatic ring	694, 537
Polyvinyl acetate (PVAc)	Asymmetric and symmetric –CH$_2$–	2920, 2853	35,39,40
	C=O ester	1737	
	Symmetric in-plane C–H(–CH$_2$)	1375 (shoulder)	
	–C–C(O)–O	1242	
Polyvinylchloride (PVC)	C–H stretching	2890–2958	29–31
	C–H$_2$ deformation	1339	
	C–H rocking	1240–1257	
	trans-C–H wagging	961	
	C–Cl stretching	844	
	cis-C–H wagging	600	

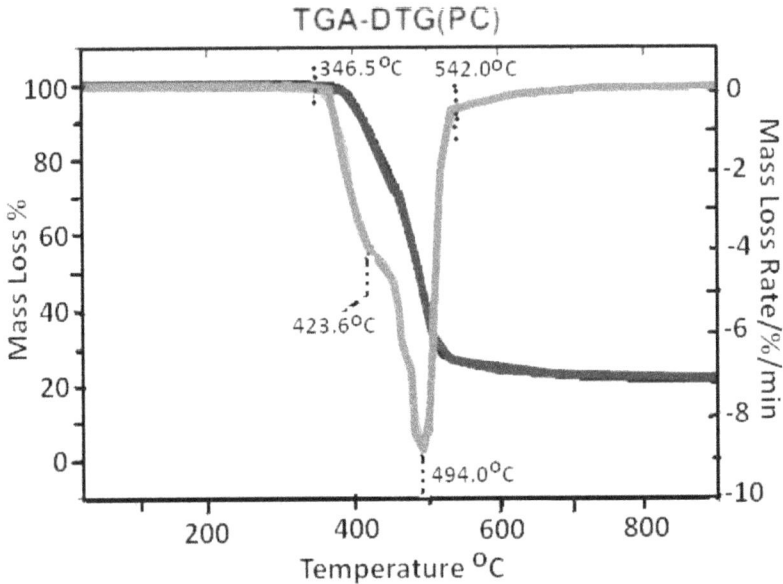

Figure 6.22.   TGA Thermogram for PC. [Reprinted with permission from X. Haung, X. Y. Ouyang, F. L. Ning, and J. Q. Wang, *Polym. Degrad. Stabil.* 91, 606 (2006). Copyright © 2006 Elsevier Ltd. All rights reserved.]

6.11. Future Trends

Polymer characterization using the tools of thermal analysis, analytical chromatography, and spectroscopy has developed within the industrial analytical research community [47]. It is important to understand the reaction pathways of polymers and their effects in order to control the process.

Instrumental analysis has long been firmly established in many industries, but the polymer industry has been less receptive. This cannot be attributed to any lack of interest or absence of problems. In the polymer industry, characterization and analysis are vitally important in connection with developing polymer technology. The hesitancy of the polymer industry is to be sought rather in the properties of the substances to be investigated. There is still a wide gap between the capabilities of present-day methods of characterization of polymers and the everyday requirements of the polymer industry.

A number of instrumental studies have been performed focusing on various properties. The application range of thermoplastic materials continues to expand due to its many useful properties, which often exceed the limits of today's commercially available polymers. Instrumental analysis can provide enhanced opportunities in a variety of applications.

References

1. S. Guilbert, B. Cuq, and N. Gontard, *Food Additives Contaminants* 14, 741 (1997).
2. K. Petersen, P. V. Nielsen, G. Bertelsen, M. Lawther, M. B. Olsen, N. H. Nilssonk, et al., *Trends Food Sci. Technol.* 10, 52 (1999).
3. J. F. Vega, A. Munoz-Escalona, A. Santamaria, and M. E. Munoz, *Macromolecules* 29, 960 (19960.
4. C. Y. Liu, J. Wang, and J. S. He, *Polymer* 43, 3811 (2002).
5. H. D. Stenzenber, *Adv. Polym. Sci.* 117, 165 (1994).
6. G. Xue, *Prog. Polym. Sci.* 22, 313 (1997).
7. I. L. Koenig, *Spectroscopy of Polymers,* American Chemical Society, Washington, DC (1992).
8. G. Xue, *Prog. Polym. Sci.* 19, 317 (1994).
9. A. C.-Y. Wong and F. Lam, *Polym. Testing* 21, 691 (2002).
10. J. V. Gulmine, P. R. Janissek, H. M. Heise, and L. Akcelrud, *Polym. Testing* 21, 557 (2002).
11. A. Riga, R. Collins, and G. Mlachak, *Thermochim. Acta* 324, 135 (1998).
12. J. P. Gibert, J. M. Lopez Cuesta, A. Bergeret, and A. Crespy, *Polym. Degrad. Stabil.* 67, 437 (2000).
13. B. Bodzay, B. B. Marosfoi, T. Igricz, K. Bocz, and G. Marosi, *J. Anal. Appl. Pyrolysis* 85, 313 (2009).
14. P. L. Ku, *Adv. Polym. Technol.* 8, 177 (1988).
15. J. La Coste, F. Delor, R. P. Singh, P. A. Vishwa, and S. Sivaram, *J. Appl. Polym. Sci.* 59, 953 (1996).
16. P. L. Ku, *Adv. Polym. Technol.* 8, 201 (1988).
17. G. Goldfinger and K. E. Lauterbach, *J. Polym. Sci.* 3, 145 (2003).
18. V. Karmore and G. Madras, *Ind. Eng. Chem. Res.* 40, 1307 (2001).
19. M. W. Beach, N. G. Rondan, R. D. Froese, B. B. Gerhart, J. D. Green, B. G. Stobby, A. G. Shmakov, V. M. Shvartsberg, and O. P. Korobeinichev, *Polym. Degrad. Stabil.* 93, 1664 (2008).
20. A. Guyot, *Polym. Degrad. Stabil.* 15, 219 (1986).
21. S. K. Bhattacharya, A. Tendokar, and A. Misra, *Mol. Cryst. Liq. Cryst.* 153, 501 (1987).
22. S. K. Sharma, A. Tendokar, and A. Misra, *Mol. Cryst. Liq. Cryst.* 157, 597 (1988).
23. S.-H. Hwanga, K.-S. Jeong, and J.-C. Jung, *Eur. Polym. J.* 35, 1439 (1999).
24. G. Colomines, A. van der Lee, J.-J. Robin, and B. Boutevin, *Eur Polym. J.* 44, 2874 (2008).
25. A. L. Andrady, in J. E. Mark (ed.), *Polymer Data Handbook,* Oxford University Press, Oxford, U.K., pp. 928–934 (1998).
26. O. G. Hansen, *PVC in the Health Care Sector, Medical Device Technology,* Octo Media, London (1995)
27. C. S. B. Nair, in *Medical Device and Diagnostic Industry, A Technical-Economic News Magazine for Medical Plastics and Pharmaceutical Industry,* Los Angeles (March 1996).

28. W. S. Penn, *PVC Technology,* 3rd ed., Applied Science, London, p. 188 (1971).
29. S. Ramesh, K. H. Leen, K. Kumutha, and A. K. Arof, *Spectrochim. Acta A* 66, 1237 (2007).
30. S. Rajendran and T. Uma, *Mater. Lett.* 44, 208 (2000).
31. S. Rajendran and T. Uma, *Mater. Lett.* 44, 242 (2000).
32. S. Rajendran and T. Uma, *J. Power Sources* 88, 282 (2000).
33. A. Marcilla, S. Garcia, and J. C. Garcia-Quesada, *Polym. Testing* 27, 221 (2008).
34. G. Madras, J. M. Smith, and B. J. McCoy, *Ind. Eng. Chem. Res.* 35, 1795 (1996).
35. P. R. Westmoreland, T. Inguilizian, and K. Rotem, *Thermochim. Acta* 367–368, 401 (2001).
36. M. Sadeghi, Gh. Khanbabaei, A. H. S. Dehaghani, M. Sadeghi, M. A. Aravand, M. Akbarzade, and S. Khatti, *J. Membr. Sci.* 322, 423 (2008).
37. G. Gozzelino and G. Malucelli, *Colloids Surf. A: Physicochem. Eng. Aspects* 235, 35 (2004).
38. A. Joseph, A. E. Mathai, and S. Thomas, *J. Membr. Sci.* 220, 13 (2003).
39. M. Hema, S. Selvasekarapandian, D. Arunkumar, A. Sakunthala, and H. Nithya, *J. Non-Crystalline Solids* 355, 84 (2009).
40. S. Rajendran, M. Sivakumar, and R. Subadevi, *Mater. Lett.* 58, 641 (2004).
41. G. Sivalingam, R. Karthik, and G. Madras, *Ind. Eng. Chem. Res.* 42, 3647 (2003).
42. M. I. Kohan, in: M. I. Kohan (ed.), *Nylon Plastics Handbook,* Hanser/Gardner, New York, p. 6 (1995).
43. K. Lamnawar, A. Baudoin, and A. Maazouz, *Eur. Polym. J.* 46, 1604 (2010).
44. S. V. Levchik and E. D. Weil, *Polym. Int.* 54, 981 (2005).
45. S. V. Levchik and E. D. Weil, *J. Fire Sci.* 24, 137 (2006).
46. X. Haung, X. Y. Ouyang, F. L. Ning, and J. Q. Wang, *Polym. Degrad. Stabil.* 91, 606 (2006).
47. S. A. Lieman, C. Phillips, W. Fitzgerald, R. A. Pesce Rodriguez, J. B. Morris, and R. A. Fifer, *ACS Symp. Ser.* 581, 12 (1994).

Chapter 7

Thermosets

Thermosetting materials, or thermosets, are multifunctional polymers that cure irreversibly upon heating. They are irreplaceable materials for many high-technology applications. After curing (cross-linking), they have excellent chemical and corrosion resistance, good thermal and mechanical stability, and superior electrical behavior. They are used particularly in high-performance structural materials in the aerospace, automotive, and electronic industries.

Thermosets are manufactured using resins that become infusible solids after completion of polymerization. Thermosets will not soften when heated. The majority of thermosetting materials are manufactured using phenol formaldehyde resins. Other thermosetting resins, such as urea formaldehyde and melamine resins, are used in certain applications. Epoxy resins of relatively low molecular weight have become familiar as thermosetting resins.

The polymerization of thermosets can be attained either by application heat or use of a chemical accelerator. Cold-setting resins such as adhesives are generally completely cured by the introduction of a chemical accelerator. However, in certain instances a minor amount of heat is essential to complete cure. Control of the processing of thermosetting materials requires accurate knowledge of the polymerization kinetics as a function of the applied processing temperature. The optimization and control procedures based on the curing and fundamental transport phenomena are intimately associated with processing technology.

Thermosetting materials require a careful application of processing conditions. The development of viscosity is dependent on the temperature and

DOI: 10.5643/9781606502440/ch7

on the polymer structure. During processing, the resin changes from a low-viscosity liquid at the starting point to a solid polymer at the end of the process. Knowledge of the curing processes of thermosetting materials can be used successfully to select the appropriate temperatures and pressures to be applied [1,2].

An understanding of the relationships between bulk properties and molecular architecture is a major goal of polymer science. Instrumental methods of analysis have been applied more and more to the study of polymers.

7.1.   Phenol Formaldehyde

Thermoset polymers are commonly referred to as network polymers or cross-linked polymers. The major portion of the thermoset polymers may be considered as one gigantic molecule. Due to cross-linking, the polymer chain mobility is limited. Thus these polymers have excellent dimensional stability as well as thermal stability.

One of the major families of thermoset polymers is phenolic polymers. Phenolic polymers always exhibit good thermal stability [3]. Major application areas are focused on flame-proof fibers, thermal insulation, and protective clothing [4] in the construction, automotive, household goods, and electrical industries [5]. Phenolic polymers are also widely used as lacquers and varnishes, molding compounds, laminates, and adhesives [6].

The major use of thermal analysis is on thermosetting polymers such as phenol formaldehyde [7], unsaturated polyesters [8], and epoxies [9]. The thermosetting material is partially cured for a known length of time in an oven. Then differential scanning calorimetry (DSC) or differential thermal analysis (DTA) is used to measure the heat required to complete the cure. The changes in the thermogram are quite marked from the uncured to the fully cured state. It is easy to detect when curing is complete or to follow the curing kinetics [10]. Thermogravimetric analysis (TGA) is based on the pyrolysis of a filled polymer to determine the amount of filler within the material [11].

The processing of thermosetting polymers involves exposure to heat at various levels. The quality of product is determined by control of the temperature distribution and the rate of temperature rise. The temperature variation during cure depends to a large extent on the heat of reaction, the specific heat, and the heat conductivity at different stages of the cure cycle. The temperature determines the quality and ultimate properties of the cured articles.

Thermal instrumental methods for studying curing do not require the presence of reinforcements [12]. The activity of catalysts and the effect of fillers or other additives on curing can be measured [13]. In the following, some representative results are given.

7.1.1.   DSC Thermogram of Phenol Formaldehyde

Differential scanning calorimetry [14–16], a method for directly measuring the reaction occurring during the cure of thermosetting systems, allows direct determination of the state of cure by measuring the amount of residual reactive material present in the sample. The polymerization of phenol and formaldehyde resins is accomplished in two steps of reaction. The addition of formaldehyde to phenol to form hydroxymethyl resorcinol is the first step, and the condensation of hydroxymethyl resorcinol to form methylene and methylene-ether bridged di- or high-molecular compounds is the second step. A typical DSC thermogram is shown in Figure 7.1 [17]. Two peaks are clearly visible, and it is generally accepted that the peaks correspond to the main reactions of formaldehyde addition and formation of ether and methylene bridges, respectively.

7.1.2.   FTIR Spectrum of Lignin–Phenol Formaldehyde Resol Resin

The FTIR spectrum of a phenol formaldehyde resol resin is illustrated in Figure 7.2 [18]. It shows that the resol contains complex functional groups with peaks at 3350, 3020, 2960, 1600, 1500, 1450, and 1100 cm^{-1}, indicating hydroxyl groups, R–H, aromatic rings, and substituted benzene, respectively.

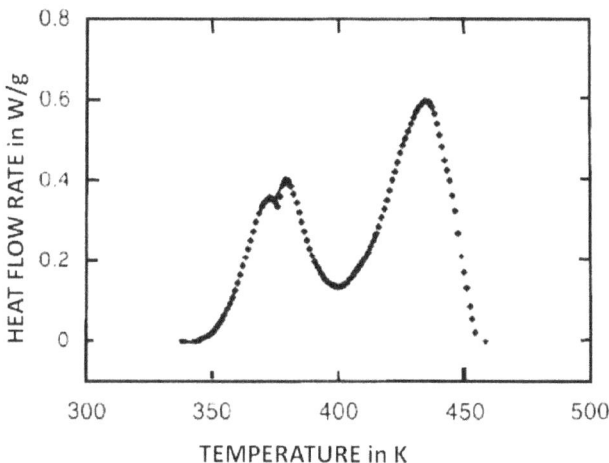

Figure 7.1.   Dynamic DSC thermogram of a phenolic system obtained at 7° C min^{-1}. [Reprinted with permission from J. M. Kenny, G. Pisaniello, F. Farina, and S. Puzziello, *Thermochim. Acta* 269/270, 201 (1995). Copyright © 1995 Elsevier B.V. All rights reserved.]

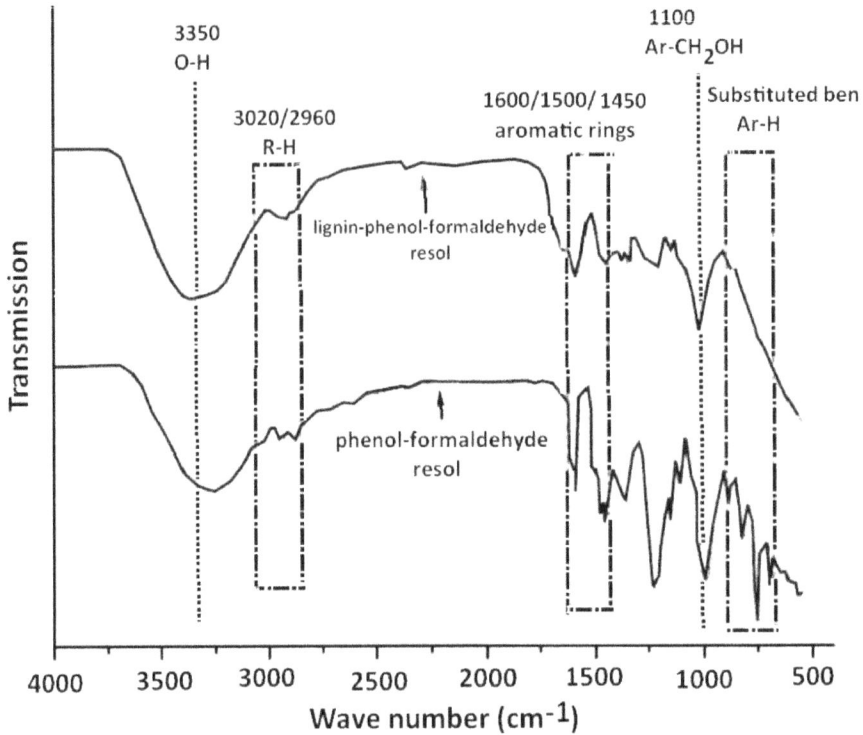

Figure 7.2. FTIR spectrum of a representative lignin–phenol formaldehyde resol resin (lignin 60%, phenol 40%). [Reprinted with permission from M. Wang, M. Leitch, and C. Xu, *Eur. Polym. J.* 45, 3380 (2009). Copyright © 2009 Elsevier Ltd. All rights reserved.]

7.1.3. DSC of Phenol Formaldehyde Resin

In differential scanning calorimetry, the residual heat of cure is measured from the area under the exothermic curve enclosed by the sigmoid baseline, which is set by computer to correct for the change in heat capacity for the sample during the transition [19]. In Figure 7.3 [20], the reference value for the total heat of cure was determined from a sample which had no precure. The initial degree of cure of this sample therefore was defined to be 0% cure or conversion. The residual heat of cure of the partially precured sample was then converted into a percent cure or degree of cure.

In this example, a positive effect of relative humidity on resin cure probably resulted from either better heat transfer because of moisture condensing on the samples when they were put into the steam treatment chamber or improved resin flow because of moisture diffusing into the samples during procuring.

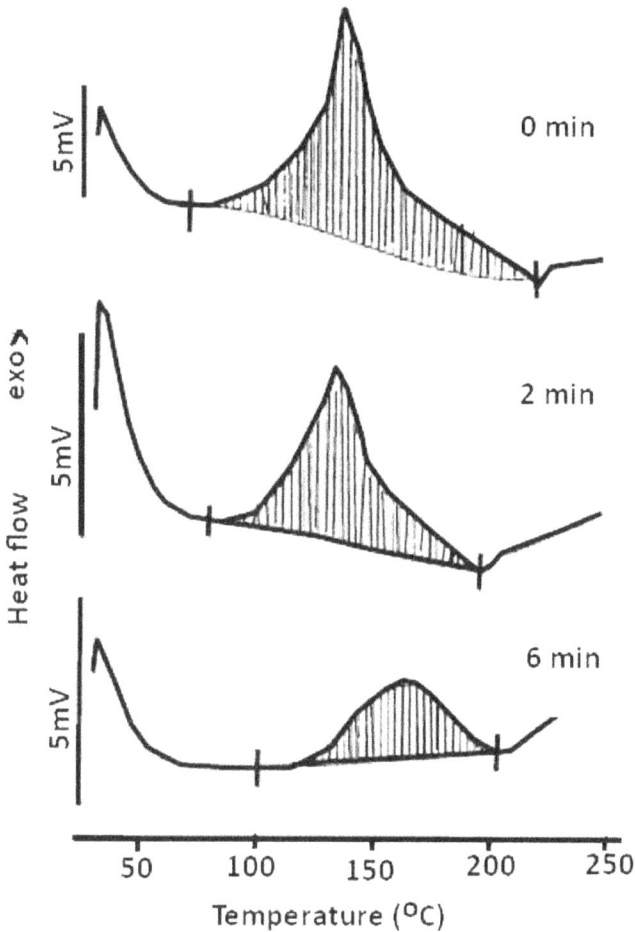

Figure 7.3.   DSC of phenol formaldehyde resin samples (at 5°C min⁻¹) after precure for various time periods at 125°C and conditioning at 89% relative humidity. [Reprinted with permission from X. M. Wang, B. Riedl, A. W. Christiansen, and R. L. Geimer, *Polymer* 35, 5685 (1994). Copyright © 1994 Elsevier Ltd. All rights reserved.]

7.1.4.   NMR Spectrum of the Phenol Formaldehyde Resin

Figure 7.4 [28] shows a typical ^{13}C NMR spectrum of a phenol formaldehyde resin with a formaldehyde/ phenol (F/P) molar ratio of 1.6/1.0. The signals are assigned to the corresponding groups using previous model compounds

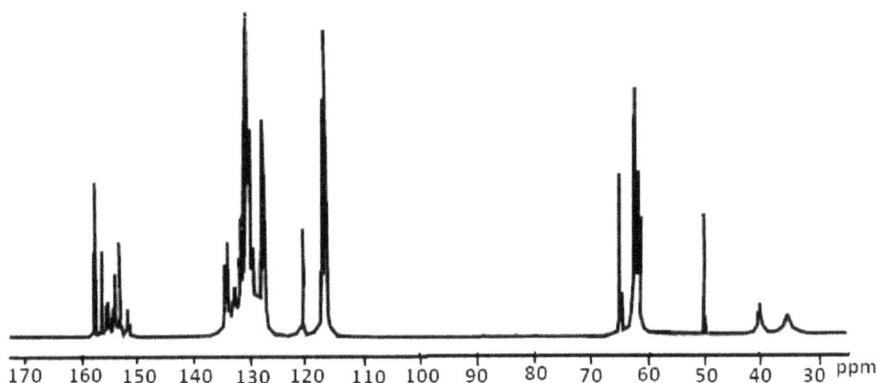

Figure 7.4. Liquid ¹³C NMR spectrum of phenol formaldehyde resin. [Reprinted with permission from D.-B. Fan, J.-Z. Li, and J.-M. Chang, *Eur. Polym. J.* 45, 2849 (2009). Copyright © 2009 Elsevier Ltd. All rights reserved.]

[21–27]. The chemical shifts are as follows: phenoxy region at 150.0–158.0 ppm; phenoxy, alkylated in para position, at 156.2–156.8 ppm; phenoxy, alkylated in ortho position, at 153.4–156.1 ppm; phenoxy, alkylated in two ortho and/or para positions, at 151.2–153.0 ppm; substituted para position at 129.0–130.4; substituted ortho position at 119.2–120.4 ppm; unsubstituted para position at 119.2–120.4 ppm; unsubstituted ortho position at 115.0–116.6 ppm; formaldehyde oligomers at 82.0–91.0 ppm; phenolic methylene ether bridges at 69.1–73.0 ppm; para methylol at 63.3–65.5 ppm; ortho methylol at 61.1–61.5 ppm; methanol at 50.1 ppm; para–para methylene bridges at 39.7–41.0 ppm; and ortho–para methylene bridges at 34.3–35.7 ppm.

7.2. Urea Formaldehyde

Urea formaldehyde resins are based on the reaction of two monomers, urea and formaldehyde. By using different conditions of reaction and preparation, an almost innumerable variety of condensed structures is possible. Urea formaldehyde resins are the most important of the so-called aminoplastic resins. Currently, approximately 6 billion tons are produced per annum worldwide, based on a usual solids content of 66% by mass.

Urea formaldehyde resins are thermosetting duromers (heavily cross-linked and therefore very inelastic polymers) and consist of linear or branched oligomeric and polymeric molecules, which also always contain some amount of monomer. Nonreacted urea is often beneficial to achieve special effects, e.g., better stability during storage. However, the presence of free formaldehyde is ambivalent. On the one hand, it is necessary to induce the hardening reaction.

After hardening, urea formaldehyde resins form an insoluble three-dimensional network which cannot be melted or thermoformed again. During processing, urea formaldehyde resins are still soluble or are dispersed in water or in the form of spray-dried powders, which in most cases are also redissolved in water for application. Many new polymeric thermoset materials have achieved well-established levels of application, and the use of urea formaldehyde has increased considerably with the advent of new instrumental methods of analysis.

7.2.1.   Proton NMR Spectrum of Urea Formaldehyde Resin

The ^{1}H NMR spectrum of urea formaldehyde resin at 60 MHz is shown in Figure 7.5 [31]. In this figure, the methoxy signal at 2.8–3.4 ppm and a shoulder on the –CONH– resonance, at 7.5–9.0 ppm had not been reported

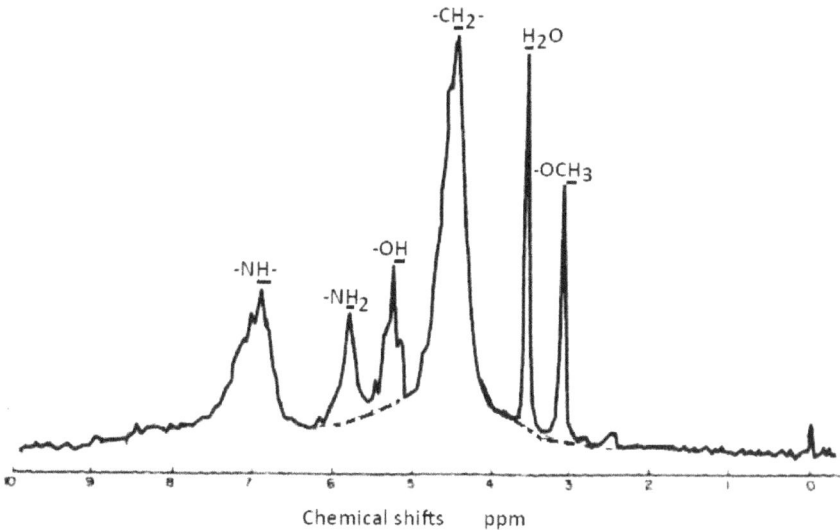

Figure 7.5.   60-MHz proton NMR spectrum of a urea formaldehyde resin. [Reprinted with permission from R. Taylor, R. J F'ragnell, J. V. Mclaren, and C. E. Snape, *Talanta* 29, 489 (1982). Copyright © 1982 Elsevier B.V. All rights reserved.]

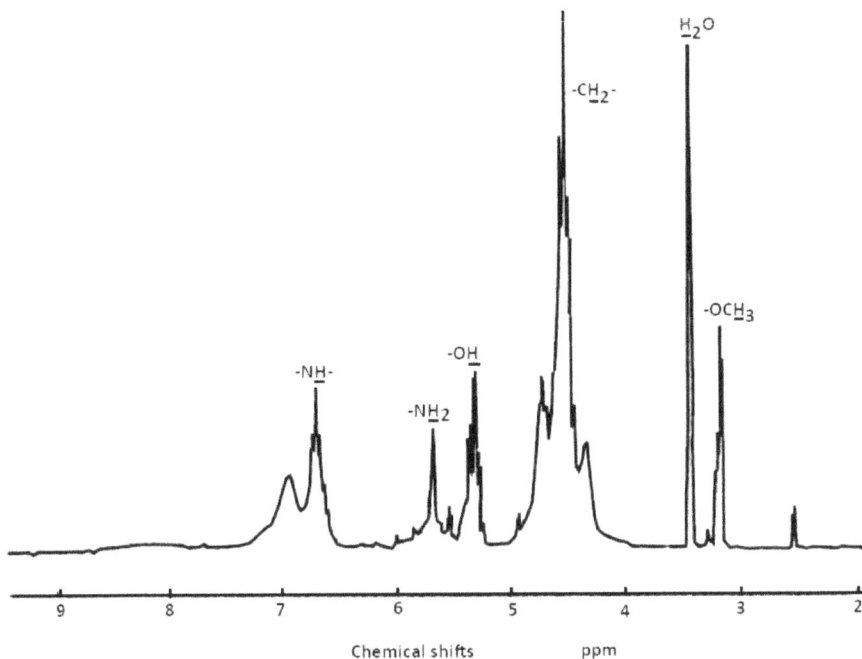

Figure 7.6. 220-proton NMR spectrum of a urea formaldehyde resin. [Reprinted with permission from R. Taylor, R. J F'ragnell, J. V. Mclaren, and C. E. Snape, *Talanta* 29, 489 (1982). Copyright © 1982 Elsevier B.V. All rights reserved.]

earlier. The methoxy signal arises because of the use of methanol-stabilized formalin in the preparation of the resin. The methanol is presumably absent from resins studied earlier [29,30].

Resolution is improved at 220 MHz (Figure 7.6) [31]. The peaks corresponding to –OH and –NH$_2$ are well resolved, and the shoulder on the –CONH– signal appears as a distinct, though very broad, peak. In addition, the breadth of the –CH$_2$– signal at 60 MHz (dotted lines in Figure 7.5) is greatly reduced at 220 MHz. All these factors result in a more accurate integration profile (see Table 7.1) [31]. At 60 MHz the width of the –CH$_2$– band results in an underestimation of the intensity of that signal and a corresponding overestimation of the –NH2, –OH, and –OCH$_3$ intensities. The higher resolution at 220 MHz of the –NH$_2$, and –OH signals improves the accuracy of the intensity calculations for both peaks.

A pulsed (FT) spectrum at 80 MHz is shown in Figure 7.7 for a much more dilute solution of urea formaldehyde resin [31]. The spectrum obtained is not as clear as that resolved at 220 MHz, but it is superior to that obtained at 60 MHz. Table 7.1 clearly shows the improvement that Fourier transform

Table 7-1 Hydrogen Distribution of a Urea Formaldehyde
Resin Measured by 1H NMR Techniques

Functional group	Chemical shift, ppm	Percentage of total hydrogen		
		60 MHz	220 MHz	80 MHz (FT)
Shoulder	7.5–9.0	5.5	3.0	3.5
Monosubstituted amide (–NH–)	6.5–7.5	21.0	23.0	22.0
Non-substituted amide (–NH$_2$–)	5.5–6.5	11.0	6.5	7.0
Hydroxyl (–OH)	5.1–5.5	11.5	8.5	10.0
Methylene (–CH$_2$–)	4.0–5.1	44.0	54.0	52.5
Methoxyl (–OCH$_3$)	2.8–3.4	7.0	5.0	5.0

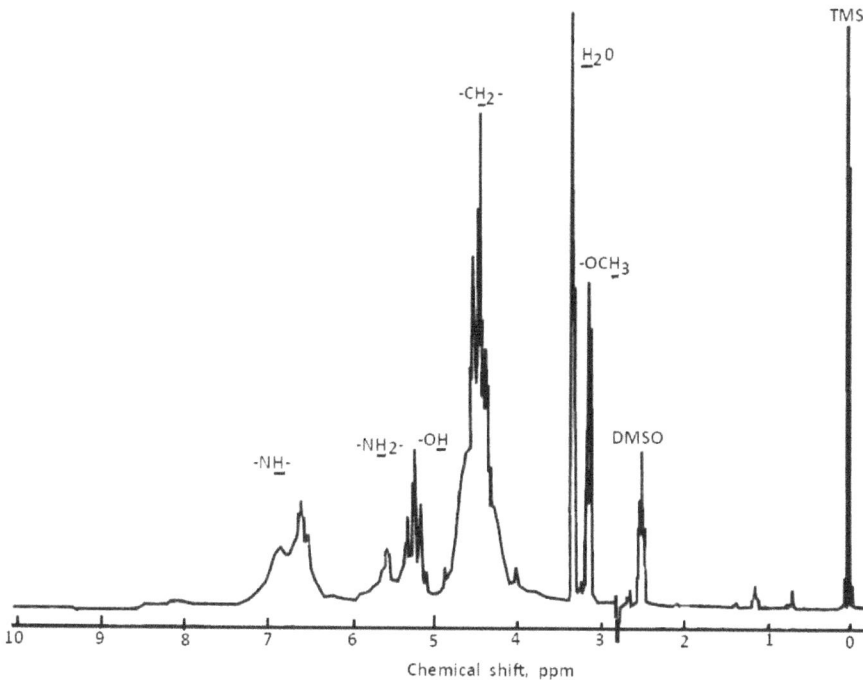

Figure 7.7. 80-MHz (FT) proton NMR spectrum of a urea formaldehyde resin. [Reprinted with permission from R. Taylor, R. J F'ragnell, J. V. Mclaren, and C. E. Snape, *Talanta* 29, 489 (1982). Copyright © 1982 Elsevier B.V. All rights reserved.]

technique provides for the low-field data, producing good agreement with results obtained at 220 MHz.

The chemical-shift range of the methylene resonances is much larger in ^{13}C NMR than in ^{1}H NMR, making it possible to obtain more detailed information on the resin structure. The ^{13}C spectrum obtained with the urea formaldehyde sample is shown in Figure 7.8 [31]. The resonances have been assigned according to Slonim et al. [32] and Tomita and Hatano. [33]. The carbon distribution obtained from the spectrum is given in Table 7.2 [31]. Two points of caution must be borne in mind when assessing the intensity of the ^{13}C signals. First, care must be taken when estimating the size of the $-CONHCH_2NHCO-$ signal because of the very close proximity of the solvent resonance bands. Second, the presence of any uron derivatives will produce an error in the estimation of the $-N(CH_2)H_2OCH_2-$ and $-NHCH_2OCH_3$ signals, owing to overlap of resonances. The presence of a carbonyl peak at 153.5 ppm, as observed with the resin investigated, indicates the presence of uron species.

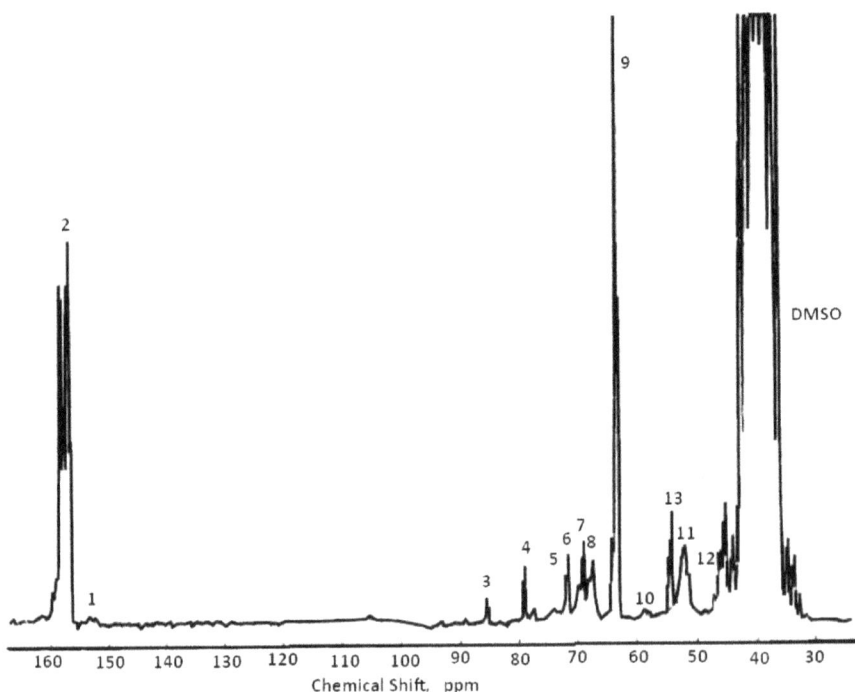

Figure 7.8. ^{13}C NMR spectrum of a urea formaldehyde resin. [Reprinted with permission from R. Taylor, R. J F'ragnell, J. V. Mclaren, and C. E. Snape, *Talanta* 29, 489 (1982). Copyright © 1982 Elsevier B.V. All rights reserved.]

Table 7-2 Data Obtained from ^{13}C NMR Spectra

Carbon type	Chemical shift, ppm	Percentage of total carbon	Concentration,* mole/100 g
Carboxyl			
1. Uron	153.5	1.0	0
2. Di-substituted urea / Tri- and tetra-substituted urea	158.6 / 157.5 – 157.7	34.5	1.0
Methylene			
3. CH_2OCH_2OH	82–87	1.0	0
4. $N(CH_2)CH_2OCH_2$	77–82	3.5	0.1
5. $N(CH_2)CH_2OCH_2$	73–77	2.5	0.1
6. $NHCH_2OCH_2$	71.5–73	2.5	0.1
7. $N(CH_2)CH_2OH$	68.5–71.5	5.0	0.1
8. $NHCH_2OCH_2$	66.5–68.5	6.0	0.2
9. $NHCH_2OH$	62–66.5	17.0	0.5
10. $N(CH_2)CH_2N(CH_2)$	57–60	1.5	0
11. $N(CH_2)CH_2NH$	49–54	11.5	0.3
12. $NHCH_2NH$	45–49	11.0	0.3
Methyl			
13. $-CH_2OCH_3$	54–55	3.0	0.1
Error		±0.5	±0.02

*Based on total urea concentration.

Source: Reprinted with permission from R. Taylor, R. J F'ragnell, J. V. Mclaren, and C. E. Snape, *Talanta* 29, 489 (1982). Copyright © 1982 Elsevier B.V. All rights reserved.

7.3. Melamine Formaldehyde

Melamine formaldehyde resins are thermosetting materials used in decorative laminates. Systematic research in this field is aimed toward improving their material properties. Attempts to improve the bulk material properties are currently based on an incomplete understanding of the final network structure of the resin. It is very likely that chemical structure, morphology, orientation, and chain dynamics can have important effects on bulk material properties. Therefore, it is important to have an understanding of the relation between the structure on the molecular level and macroscopic behavior. However, the ability to analyze the network structure of cured melamine formaldehyde resins has been far outdistanced by the rapid commercial development of the resin technology. Limited solubility of cured melamine formaldehyde resins, limiting the use of liquid chromatography and liquid NMR, and the complex chemical structure are the main reasons for this. The biggest advantage of Raman spectroscopy is the ease and versatility of sampling, so that solid melamine formaldehyde resins can be studied on a micrometer scale or in situ during curing.

Melamine formaldehyde resin formation consists of two stages [34]. During the first stage the water-insoluble melamine is dissolved in formalin and a series of addition and condensation reactions takes place to give a low-molecular-weight resin. At this stage, the resin is both soluble and fusible. During the second stage, the resin is cured with the application of heat or an acid catalyst. During curing, further chain extension and cross-linking take place to form an insoluble, infusible, three-dimensional network. The addition of formaldehyde ($HOCH_2OH$ in water) to melamine takes place under slightly alkaline conditions. Nine distinct methylolmelamines (MMs), from mono- to hexamethylolmelamine, are formed in a complex series of competitive and consecutive equilibria [35].

7.3.1.   FT Raman Spectra of Melamine Formaldehyde Resin

In the Raman spectrum of melamine, intense, sharp bands at 676 and 975 cm^{-1} can be observed, as shown in Figure 7.9a [36]. Both these bands have been attributed to deformations of the triazine ring [37]. Surprisingly, the 676 cm^{-1} band is almost absent in the spectrum of melamine formaldehyde cured in Figure 7.9b [36].

7.4.   Epoxy Thermosets

Epoxy resins are used as a matrix in a large number of polymer-matrix composites because of the large number of compounds that can react with the epoxy ring to form resin systems with a very wide range of properties [38]. Epoxies are being used in fiber-reinforced polymer (FRP) composites for various applications, e.g., automobiles, ships, sport, aerospace, and windmill blades. During service, they may be subject to different kinds of loading, either static or dynamic, in a wide range of loading rates, e.g., sport equipment at high loading rate to pressure vessels at low loading rate.

Epoxy is by far the most widely used polymer matrix for structural composites. This is due to the strong adhesiveness of epoxy, in addition to the

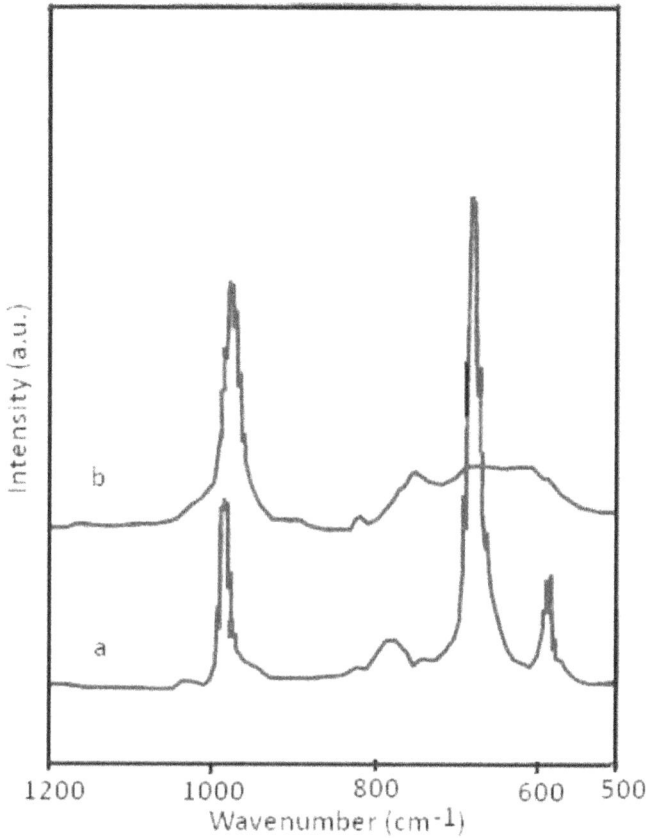

Figure 7.9.   FT Raman spectra from 500 to 1800 cm⁻¹ of (a) melamine and
(b) a cured melamine formaldehyde resin. [Reprinted with permission from
M. L. Scheepers, R. J. Meier, L. Markwort, J. M. Gelan, D. J. Vanderzande,
and B. J. Kip, *Vibrat. Spectrosc.* 9, 139 (1995). Copyright © 1995 Elsevier
B.V. All rights reserved.]

long history of its use in composites. Epoxy displays an excellent combination of mechanical properties and corrosion resistance, is dimensionally
stable, exhibits good adhesion, and is relatively inexpensive. Moreover, the
low molecular weight of uncured epoxide resin in the liquid state results in
exceptionally high molecular mobility during processing.

This mobility helps the resin to spread quickly on the surface of carbon
fiber, for example. Epoxy resins are characterized by having two or more
epoxide groups per molecule. An epoxy is a thermosetting polymer that
cures upon mixing with a catalyst (also known as a hardener). This curing
process is a reaction that involves polymerization and cross-linking.

7.4.1   Pyrogram of Cross-Linked Epoxy Resin (Printed Circuit Board)

The epoxy resins are distinguished, among thermosetting resins, by ease of handling, thermal stability, chemical resistivity, and their outstanding mechanical and electrical properties. The preparation of epoxy resins is performed normally by reaction of polyphenols, preferably diphenols of the type represented by 4,4'-dihydroxydiphenylmethane (bisphenol A), with epichlorohydrin. This reaction yields the diglycidyl ether, or low-molecular-weight polymers, of bisphenol A which have epoxide end groups.

Printed circuit board (PCB) is a thermoset material used in personal computer motherboards, among other applications. PCBs are usually made of cross-linked epoxy resin that is a thermoset of bisphenol A diglycidyl ether. Figure 7.10 shows a program of a PCB [39]. The major pyrolysates detected are phenol, methylphenol, methylethylphenol, bromophenol, dibromophenol, and various bromine-substituted bisphenol As, indicating that brominated bisphenol A is used along with bisphenol A as part of the

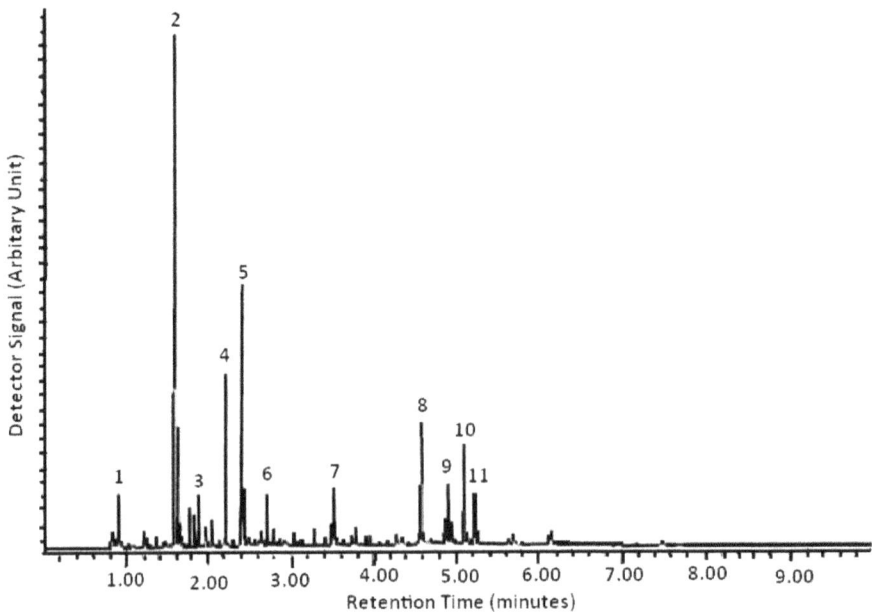

Figure 7.10.   Pyrogram of a printed circuit board epoxy polymer. [Reprinted with permission from F. C. Y. Wang, *J. Chromatogr. A* 886, 225 (2000). Copyright © 2000 Elsevier B.V. All rights reserved.]

monomer unit in the polymer. The tentative identification of major pyroly-sates from Figure 7.10 is (1) acetone, (2) phenol, (3) 2-bromophenol, (4) 4-methylethylphenol, (5) 4-methylethenylphenol, (6) 2,6-dibromophenol, (7) 2,6-dibromo-4-methylethenylphenol, (8) bisphenol A, (9) bromobisphe-nol A, (10) dibromobisphenol A, and (11) tribromobisphenol A.

7.5.   Future Trends

In order to characterize the thermosets, knowledge about the free constitu-ents is important. In particular, for thermosets with low ratios of ingredients, the concentration of free ingredient can be significant. This might have con-sequences for the network structure, the shelf-life of resins, and the proper-ties of the cured resin.

Instrumental methods have been used extensively in research on ther-moset materials. Knowing the properties and parameters of thermosets have become key to understanding the materials' long-term behavior. The instru-mental methods normally use some kind of approach that simplifies the char-acterization of the material.

References

1. J. M. Kenny, A. Apicella, and L. Nicolais, *Polym. Eng. Sci.* 29, 973 (1989).
2. J. M. Kenny, A. M. Maffezzoli, and L. Nicolais, *Composite Sci. Technol.* 38, 339 (1990).
3. A. Knop and L. A. Pilato, *Phenolic Resins,* Springer-Verlag, Berlin, p. 140 (1985).
4. J. Economy, L. C. Wohrer, F. J. Frechette, and G. Y. Lei, *Appl. Polym. Symp.* 21, 81 (1973).
5. J. F. Keegan, in I. I. Robin (ed.), *Handbook of Plastic Materials and Technology,* John Wiley, New York, pp. 181-204 (1990).
6. M. P. Stevens, *Polymer Chemistry,* 3rd ed., Oxford University Press, New York, p. 471 (1990).
7. R. H. White and T. R. Rust, *J. Appl. Polym. Sci.* 9, 777 (1965).
8. M. R. Kamal and S. Sourour, *Polym. Eng. Sci.* 13, 59 (1973).
9. D. A. Gibboney, *Soc. Plastics Eng. Tech. Paper 18,* 224 (1972).
10. L. W. Crane, P. J. Dynes, and D. H. Kaelble, *J. Polym. Sci.: Polym. Lett. Ed.* 11, 533 (1973).
11. J. Chiu, *Appl. Polym. Symp.* 2, 25 (1966).
12. Yu. V. Maksimov, V. S. Gorshkov, K. H. Melevskaya, and R. G. Krylova, *Sb. Tr. Vses. Nauch.-Issled. Inst. Nov. Stroit. Muter.* No. 25, 94 (1969); *Chem. Abstr.* 75, 77578.

13. A. R. Westwood, in H. G. Wiedemann (ed.), *3rd Thermal Anal. Proc. Int. Conf. 1971, Vol. 3,* Burkhauser, Basel (1972).

14. K. Horie, H. Hiura, M. Sawada, I. Mita, and H. Kambe, *J. Polym. Sci. A-1* 8, 1357 (1970).

15. M. A. Acitelli, R. B. Prime, and E. Sacher, *Polymer* 12, 335 (1971).

16. S. Y. Shoy, *SPE J.* 26, 51 (1970).

17. J. M. Kenny, G. Pisaniello, F. Farina, and S. Puzziello, *Thermochim. Acta* 269/270, 201 (1995).

18. M. Wang, M. Leitch, and C. Xu, *Eur. Polym. J.* 45, 3380 (2009).

19. R. A. Follensbee, J. A. Koutsky, A. W. Christiansen, G. E. Myers, and R. L. Geimer, *J. Appl. Polym. Sci.* 47, 1481 (1993).

20. X. M. Wang, B. Riedl, A. W. Christiansen, and R. L. Geimer, *Polymer* 35, 5685 (1994).

21. C. Zhao, A. Pizzi, and S. Garnier, *J. Appl. Polym. Sci.* 74, 359 (1999).

22. G. B. He and B. Riedl, *J. Polym. Sci. B* 41, 1929 (2003).

23. B. D. Park and L. B. Ried, *J. Appl. Polym. Sci.* 77, 841 (2000).

24. R. Smit, A. Pizzi, C. J. Schutte, and S. O. Paul, *J. Macromol. Sci. A* 26, 825 (1989).

25. M. G. Kim, L. W. Amos, and E. Barnes, *Ind. Eng. Chem. Res.* 29, 2032 (1990).

26. T. Holopainen, L. Alvila, J. Rainio, and T. T. Pakkanen, *J. Appl. Polym. Sci.* 66, 1183 (1997).

27. P. Luukko, L. Alvila, T. Holopainen, J. Rainio, and T. T. Pakkanen, *J. Appl. Polym. Sci.* 82, 258 (2001).

28. D.-B. Fan, J.-Z. Li, and J.-M. Chang, *Eur. Polym. J.* 45, 2849 (2009).

29. S. Kambanis and R. C. Vasishth, *J. Appl. Polym. Sci.* 15, 1911 (1971).

30. M. Chiavarini, N. Del Fanti, and R. Bigatto, *Angew. Makromol. Chem.* 46, 151 (1975).

31. R. Taylor, R. J F'ragnell, J. V. Mclaren, and C. E. Snape, *Talanta* 29, 489 (1982).

32. Ya. Slonim, S. G. Alekseveya, Ya. G. Urman, B. M. Arshava, B. Ya. Akselrod, and I. M. Gurman, *Polym. Sci. USSR* 19, 899 (1977).

33. B. Tomita and S. Hatono, *J. Polym. Sci.* 16, 2509 (1978).

34. J. R. Ebdon, B. J. Hunt, and M. Al-Kinany, *Spec. Publ. R. Soc. Chem.* 87, 109 (1991).

35. M. Gordon, A. Halliwell, and T. Wilson, *The Chemistry of Polymerization Processes,* SCI Monograph No. 20, Society of Chemical Industry, London, p. 1 (1965).

36. M. L. Scheepers, R. J. Meier, L. Markwort, J. M. Gelan, D. J. Vanderzande, and B. J. Kip, *Vibrat. Spectrosc.* 9, 139 (1995).

37. F. R. Dollish, W. G. Fateley, and F. F. Bentley, *Characteristic Raman Frequencies of Organic Compounds,* John Wiley, New York (1973).

38. J. M. Margolis, *Advanced Thermoset Composites—Industrial and Commercial Applications,* Van Nostrand Reinhold, New York (1985).

39. F. C. Y. Wang, *J. Chromatogr. A* 886, 225 (2000).

Chapter 8

Polymer Blends and Composites

Most polymers are incompatible with one another; therefore most binary polymer blends are two-phase systems. The properties of such blends are functions not only of the chemical structure of the polymers, but also of the phase morphology. The blends usually assume a reasonably fine morphology when incompatible components are mixed mechanically.

Polymer blends and composites are widely used in science and industry. The spectrum of properties blends and composites is wider than that of pure polymers. There is an economic aspect also: Sometimes fillers of polymers are inexpensive, so the output of manufacture is higher per employed pure material if fillers or blends are used.

Most industrially important polymers are in fact multicomponent systems, including polymer blends and copolymers. Their macroscopic properties are determined by microscopic and molecular factors such as the degree of phase separation, the morphology, domain sizes, interfaces, and the composition of the different phases.

8.1. Polymer Blends

Blending two polymers with different molecular structures or molecular weights may improve rheological, chemical, mechanical, and physical properties [1–3]. Polymer blends provide an economic incentive for synthesizing

DOI: 10.5643/9781606502440/ch8

new materials. Blending waste polymers has economic and ecological savings due to recycling of mixed waste plastics [4–8].

The recycling of polymer waste blends is attractive because of the elimination of a separation step. It gains significant attraction by reuse of polymeric materials. Blends of low-cost materials such as polyethylene, polypropylene, polystyrene, and polyvinylchloride comprise a significant volume of polymeric material waste.

The use of polymer blends has intensified in many technological products, due to their scientific interest and usefulness. Polymer blends are, by definition, physical mixtures of structurally different polymers. Blending can generate a material with unique properties and/or processability. It allows quick modification of performance, since most polymer blends are processed by standard processing equipment, and blends provide several other advantages as well. Some interesting polymer blend characterizations are described below.

8.1.1.   TGA of PVC and Liquid Natural Rubber

A comparison of thermogravimetric traces of polyvinylchloride (PVC), liquid natural rubber (LNR), epoxidized LNR-20, and epoxidized LNR-50 are shown in Figure 8.1. A two-stage degradation is seen with PVC in Figure 8.1a [10]. The first stage begins at 180°C and ends at 375°C, with a peak at 300°C. This corresponds to a weight loss of 59.5%, which is attributed to the elimination of HCl molecules, leaving behind longer polyene chains. The second-stage degradation begins at 375°C and ends at 490°C, with a peak temperature at 466°C. Thermal degradation of polyene sequences occurs and yields volatile aromatic and aliphatic compounds by the intramolecular cyclization of the conjugated sequences [9]. The total weight loss at this stage is 90%.

In Figure 8.1b, liquid natural rubber exhibits only a single-stage degradation, which occurs between 240 and 450°C. The peak temperature is observed at 384°C. The weight loss at the end is 98%. Fragmentation of polyisoprene chains must occur during the degradation. This yields volatile fragments such as isoprene and dipentene [11,12].

The degradation of epoxidized liquid natural rubber (LNR)-20 and epoxidized LNR-50 follows a similar pattern, as shown in Figures 8.1c and 8.1d. These also show only single-stage degradation. The onset temperatures are from 240 to 275°C and to 295°C, respectively. With increases in epoxy content from 0 to 20 and then 50 mol%, the peak temperatures increase from 384 to 396 and to 415°C, respectively. Weight losses for epoxidized LNR-20 and

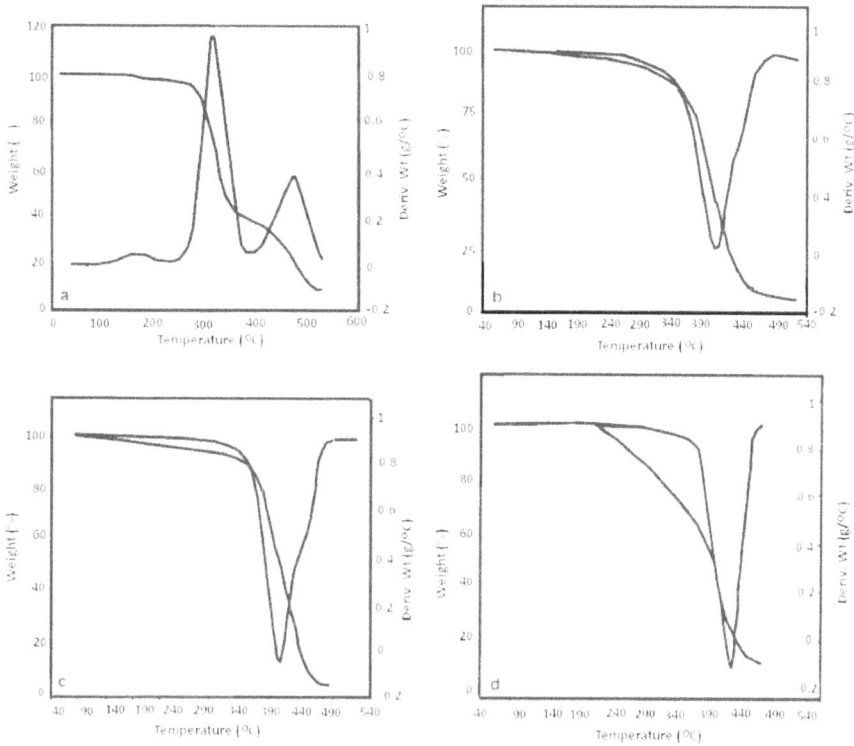

Figure 8.1. Thermogravimetric curves of (a) polyvinylchloride, (b) liquid natural rubber (LNR), (c) epoxidized LNR-20, and (d) epoxidized LNR-50. [Reprinted with permission from M. N. Radhakrishnan Nair, G. V. Thomas, and M. R. Gopinathan Nair, *Polym. Degrad. Stability* 92, 189 (2007). Copyright © 2007 Elsevier Ltd. All rights reserved.]

epoxidized LNR-50 are 96% and 93%, respectively. The experiment indicates that the rubber molecules delay the breakdown of the molecules due to the presence of the epoxy group. The more polar epoxy groups lead to higher interaction among the rubber molecules.

8.1.2.   Raman Spectra of Phenoxy PMMA Blends

Phenoxy is a simplified term for poly(hydroxy ether of bisphenol A). Figure 8.2 [13] illustrates the Raman spectra of pure phenoxy and pure polymethylmethacrylate (PMMA). Distinctive differences between PMMA and phenoxy can be observed from their Raman spectra in the 1550–1800 cm^{-1} region. Pure PMMA shows a Raman band at 1730 cm^{-1}. It is attributed to the

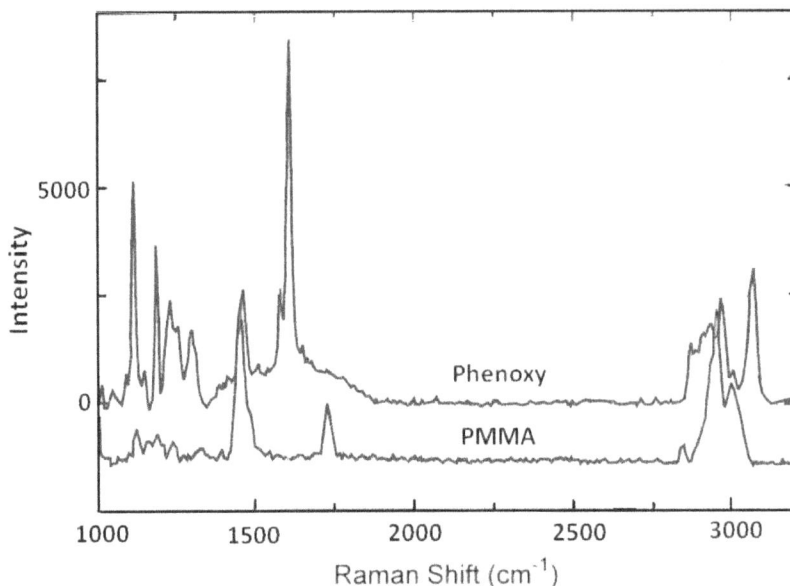

Figure 8.2. Raman spectra of PMMA and phenoxy obtained in the range 1000–3200 cm^{-1}. [Reprinted with permission from Y. Ward and Y. Mi, *Polymer* 40, 2465 (1999). Copyright © 1999 Elsevier Science Ltd. All rights reserved.]

stretching mode of carbonyl groups. Pure phenoxy shows a band at 1610 cm^{-1} that is ascribed to the stretching mode of the phenyl rings.

Blends of phenoxy and PMMA show Raman bands at 1610 and 1730 cm^{-1}. In Figure 8.3 [13], the intensity ratio of Raman bands is a function of the composition in the blends. The advantage of Raman technique is particularly phase separation and its microsampling capability.

The increased chain mobility at higher temperature makes it possible to form hydrogen bonds between the hydroxyl group of phenoxy and the carbonyl group in PMMA. Above the glass transition temperature (T_g) of the polymer, there is sufficient energy to disrupt the self-associated hydroxyl groups in phenoxy. Chain mobility is improved with functional groups. They are more accessible for participating in the intermolecular interactions.

8.1.3. DSC of Polypropylene and Low-Density Polyethylene

For the binary blend of polypropylene (PP) and low-density polyethylene (LDPE) of a specific blend ratio of 90:10% w/w, two well-separated peaks

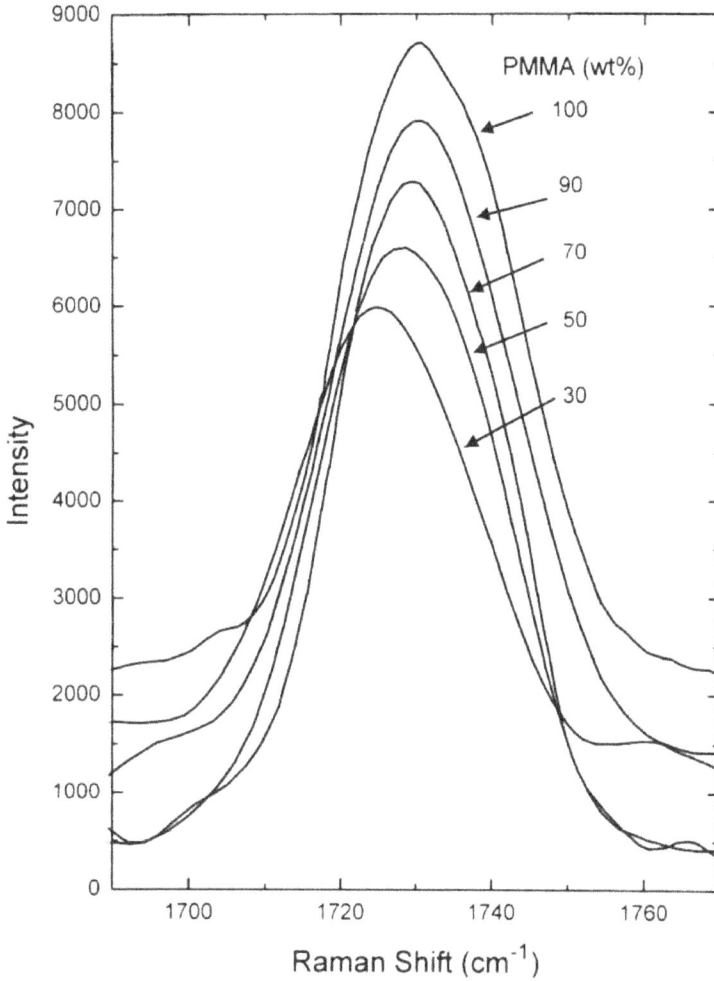

Figure 8.3. Raman spectra of phenoxy/PMMA blends with different compositions in the carbonyl stretching region. [Reprinted with permission from Y. Ward and Y. Mi, *Polymer* 40, 2465 (1999). Copyright © 1999 Elsevier Science Ltd. All rights reserved.]

are obtained on the DSC thermogram as shown in Figure 8.4 [14]. The peak temperatures are referred to the melting temperatures of the blending components. The lower melting-point peak corresponds to the LDPE and the higher melting-point peak corresponds to the PP. The blend of PP and LDPE is basically incompatible regardless of the molecular structure of PE. A compatible blend will have single peak on the thermogram. The incompatibility

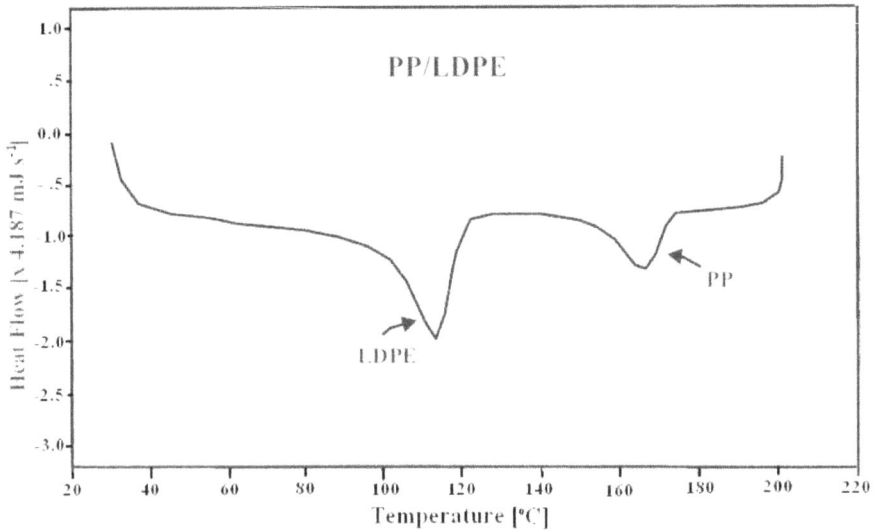

Figure 8.4. Typical double-peak DSC thermogram for blends of poly-propylene and low density polyethylene. [Reprint with permission from A. C.-Y.Wang and F. Lam, *Polym. Testing* 21, 691 (2002). Copyright © 2002 Elsevier Science Ltd. All rights reserved.]

here might be due to the difference in viscosity between the blending components [15].

8.1.4. FTIR of a Nylon 6 Blend

The assignments of various infrared absorption bands of nylon 6 are well established [16–18]. The spectrum of solvent-cast nylon 6 has strong absorption bands at 3299 cm^{-1} due to hydrogen-bonded NH stretching, at 2943 and 2869 cm^{-1} due to CH stretching of the methylene group, at 1641 cm^{-1} due to the amide I band (predominantly C=O stretching), and at 1546 cm^{-1} due to the amide II band (C–N stretching and NH deformation).

As shown in Figure 8.5 [19], the spectrum of a cast film of poly(ethylene-vinyl alcohol) shows strong absorption bands at 3450 and 1115 cm^{-1} due to the hydroxyl stretch alcohol group; at 1720 and 1190 cm^{-1}, the characteristic of the alcohol group; and at 1720 and 1190 cm^{-1}, the characteristic frequencies of a formate group [20,21]. The infrared spectra of poly(ethylene-vinyl alcohol) suggests that there is partial conversion of the alcohol groups into formate groups. This is perhaps a consequence of the hydrolysis of the ester

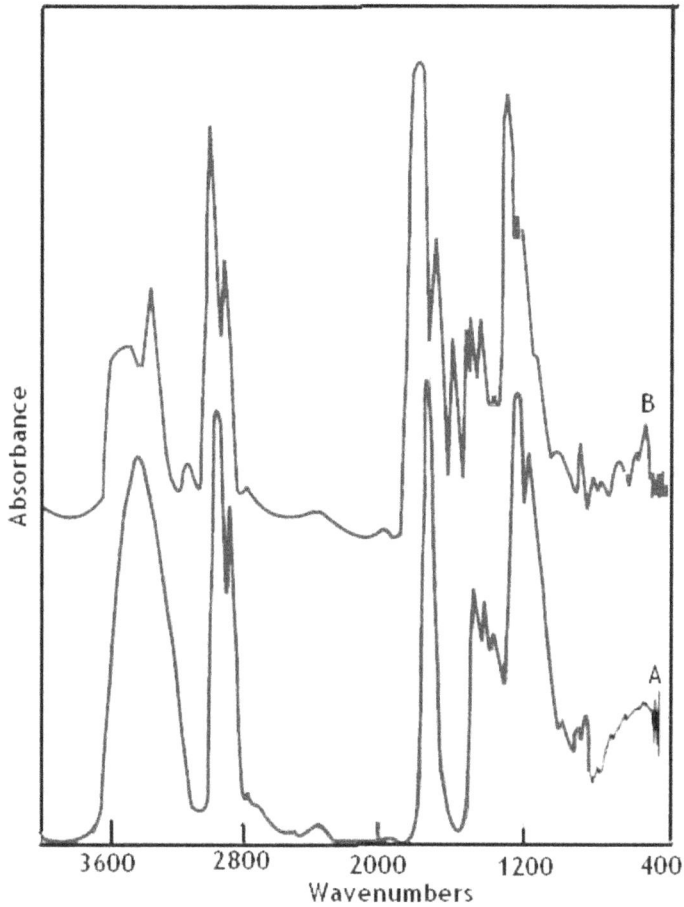

Figure 8.5. FTIR spectra of formic acid cast films: (A) pure poly(ethylene-vinyl alcohol) and (B) 25/75 nylon-copolymer blend. [Reprinted with permission from G. M. Venkatesh, R. D. Gilbert, and R. E. Fornes, *Polymer* 26, 45 (1985). Copyright © 1985 Elsevier Ltd. All rights reserved.]

groups formed when dissolved in formic acid, back to alcohol in the presence of acid [22].

8.1.5. IR Spectrum of Nylon 4,6 and Ethylvinylalcohol (EVOH)

Figure 8.6 [23] shows the relative intensity of the IR peak around the 1700 cm⁻¹ region for a blend of nylon 4,6 and ethylvinylalcohol (EVOH). The

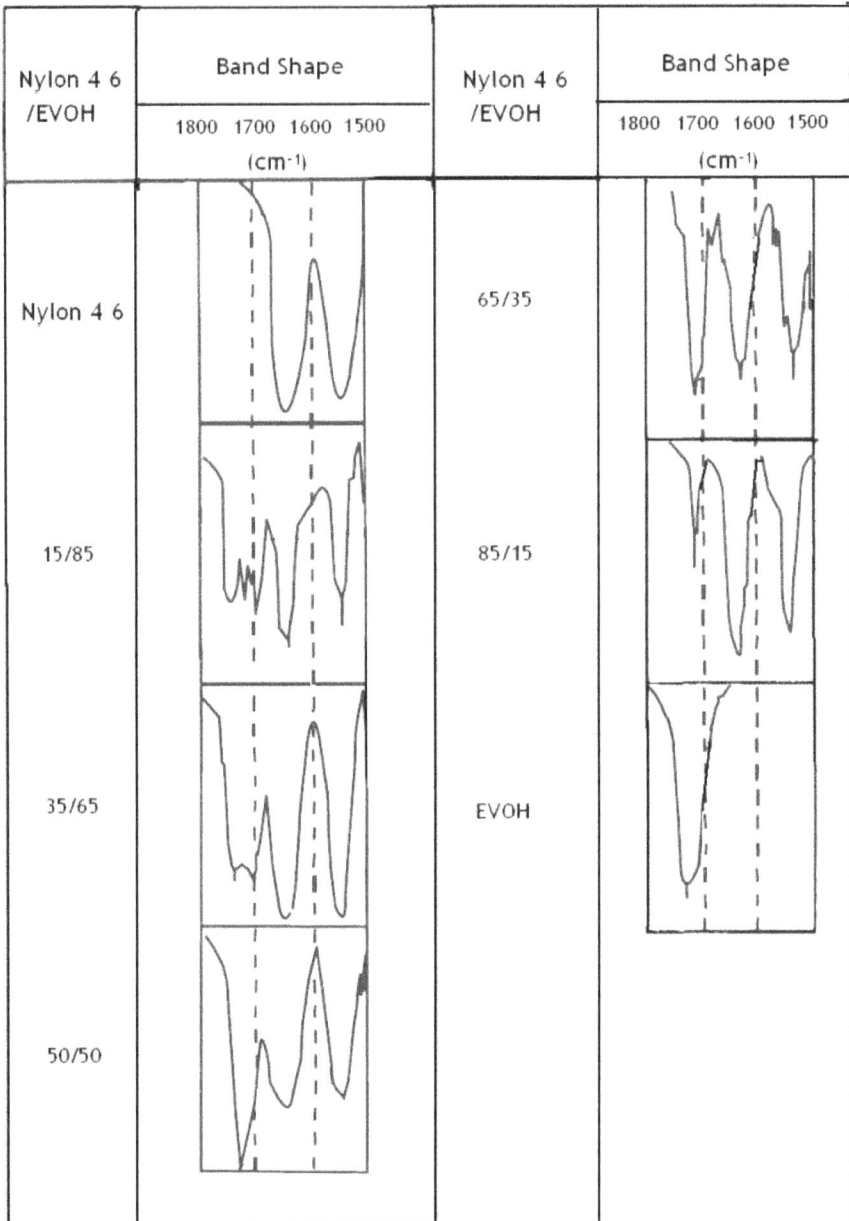

Figure 8.6. Relative intensity of the peak around 1700 cm⁻¹ of nylon 4,6/EVOH blends on IR spectrum as a function of blend concentrations. [Reprinted with permission from C.-S. Ha, M.-G. Ko, and W.-J. Cho, *Polymer* 38, 1243 (1997). Copyright © 1997 Elsevier Ltd. All rights reserved.]

peaks of the carbonyl group in EVOH shift to lower wave number up to 35% of nylon 4,6 content by 10 cm^{-1}, but no shift in the peaks is observed for the blends having more than 65% nylon 4,6.

The peak shift suggests that there exists some interaction between EVOH and nylon at higher EVOH contents. The interaction is also evidenced with 85% and 65% EVOH, indicating a strong interaction. When the composition of the blend is more than 35% nylon 4,6, however, the interaction between the components becomes weaker with increasing nylon 4,6 concentration. This suggests that a certain molecular interaction exists between the N–H group of nylon 4,6 and the C–O group of EVOH. It is suggested that these interactions are hydrogen bonding and/or dipole-–dipole interactions between the C–O group of EVOH and the N–H group in nylon4,6, and that these interactions may vary with the concentration of the blend.

When the concentration of nylon 4,6 is higher, the interaction is thought to be lower, because of strong intramolecular interactions in each homopolymer such as hydrogen-bonding characteristics resulting in preventing the intermolecular interaction between nylon 4,6 and EVOH.

8.2.   Polymer Composites

Polymer science has entered an era of great emphasis on the technological applications of existing polymers. The art of polymer composites has come to mean the mixing of two or more polymers and reinforcing materials with differing properties to obtain synergistic effects. The use of composites to develop specific properties in plastics has grown enormously in practical importance, and a considerable amount of research has been devoted to understanding the principles of property development in composite systems.

Composite materials are created by combining two or more components to achieve desired properties. For example, wood-based composites using thermoplastics as a continuous phase can result in better water resistance and dimensional stability compared to composites with low polymer content [24]. Wood composites are manufactured using conventional thermoplastic processing equipment [25], and they can be used to replace impregnated wood in many outdoor applications.

Composite materials have gained significant importance in the design of new systems. For many applications, information is needed for characterization of materials. Identifying and characterizing composite materials is crucial in a number of industrial processes. For instance, temperature parameters for composite materials cannot be determined unless a thermal analysis is available.

Despite the importance of these material properties, characterization of composite materials is only partially understood. The effective character-ization of a composite material is a complex function of the experimental approach to determine the precise value of a particular parameter. Some results for analysis of polymer composite materials by instrumental methods are given below.

8.2.1.   Measurement of Torque on Polypropylene/ Wood Flour/Modified Polypropylene

Composites containing 30 wt% wood flour and up to 10 wt% of functional-ized polypropylene (PP) produce a typical mixing torque profile as shown in Figure 8.7 [26]. The initial loading peak reflects the high viscosity of unmelted PP. Decreasing loading peak is due to the melting of the PP. Fur-ther addition of components again produces an increase of the torque. The decrease again shows due a slight decrease as the coupling agent melts and filler is dispersed in the polymer matrix. The torque is stabilized once the dispersion is completed, after which the torque value remains nearly constant. The changes in the composition of the systems do not affect the torque values.

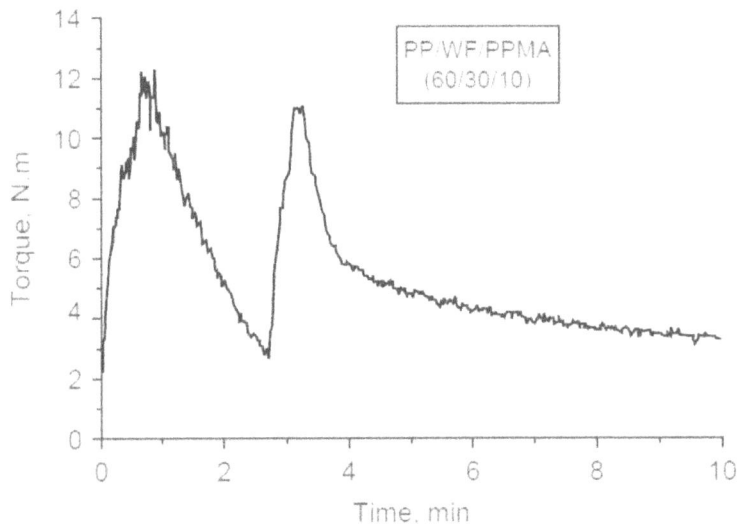

Figure 8.7.   Mixing torque profile of a composite. [Reprinted with permis-sion from S. M. B. Nachtigall, G. S. Cerveira, and S. M. L. Rosa, *Polym. Testing* 26, 619 (2007). Copyright © 2007 Elsevier Ltd. All rights reserved.]

8.2.2. DSC of Wood/HDPE

The dehydration of a wood/high-density polyethylene (HDPE) composite can be confirmed by its differential scanning calorimetry curve. Figure 8.8 [27] illustrates such a curve from 30 to 170°C. Air circulation accelerates the dehydration. The moisture content exposed to air and nitrogen is the same. Therefore, at 37°C, the effect of oxygen on the degradation of the composite is not significant.

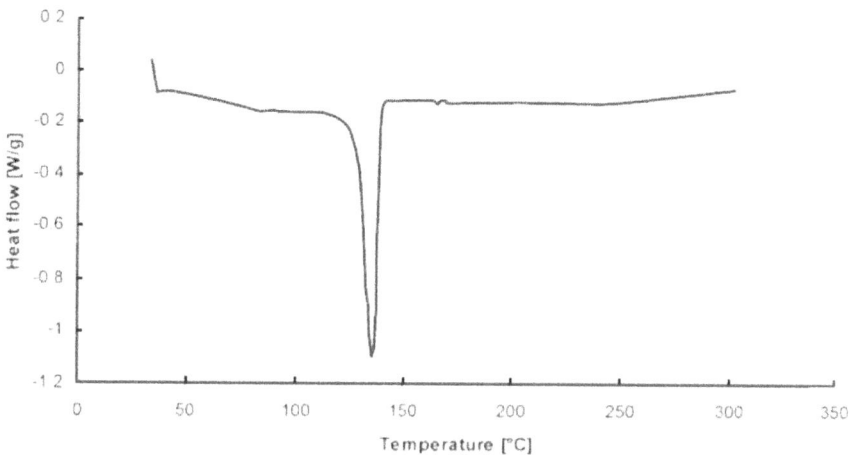

Figure 8.8. DSC curve of a wood/HDPE composite. [Reprinted with permission from R. Li, *Polym. Degrad. Stability* 70, 135 (2000). Copyright © 2000 Elsevier Ltd. All rights reserved.]

8.2.3. FTIR of Lignocellulosic Reinforcement and HDPE Matrix

Figure 8.9 [29] shows the characteristic spectra of composites containing 40% lignocellulosic fibers. It shows the bands involved in the interfacial interaction between the lignocellulosic reinforcement and the HDPE matrix, in the presence or absence of different types of coupling agents. Simulated spectra are obtained by means of the weighted spectral addition of each component of the composite. By comparing the composites' spectra with the simulated ones, differences in the shape and intensity of the bands can be observed.

The most significant bands of the lignocellulosic fibers are C=O stretching of acetyl or carboxylic acid at 1740 cm^{-1}, absorbed H-O at 1653/1635 cm^{-1}, aromatic bending of C–H (ring) at 1507 cm^{-1}, lignin and CH$_2$ symmetric

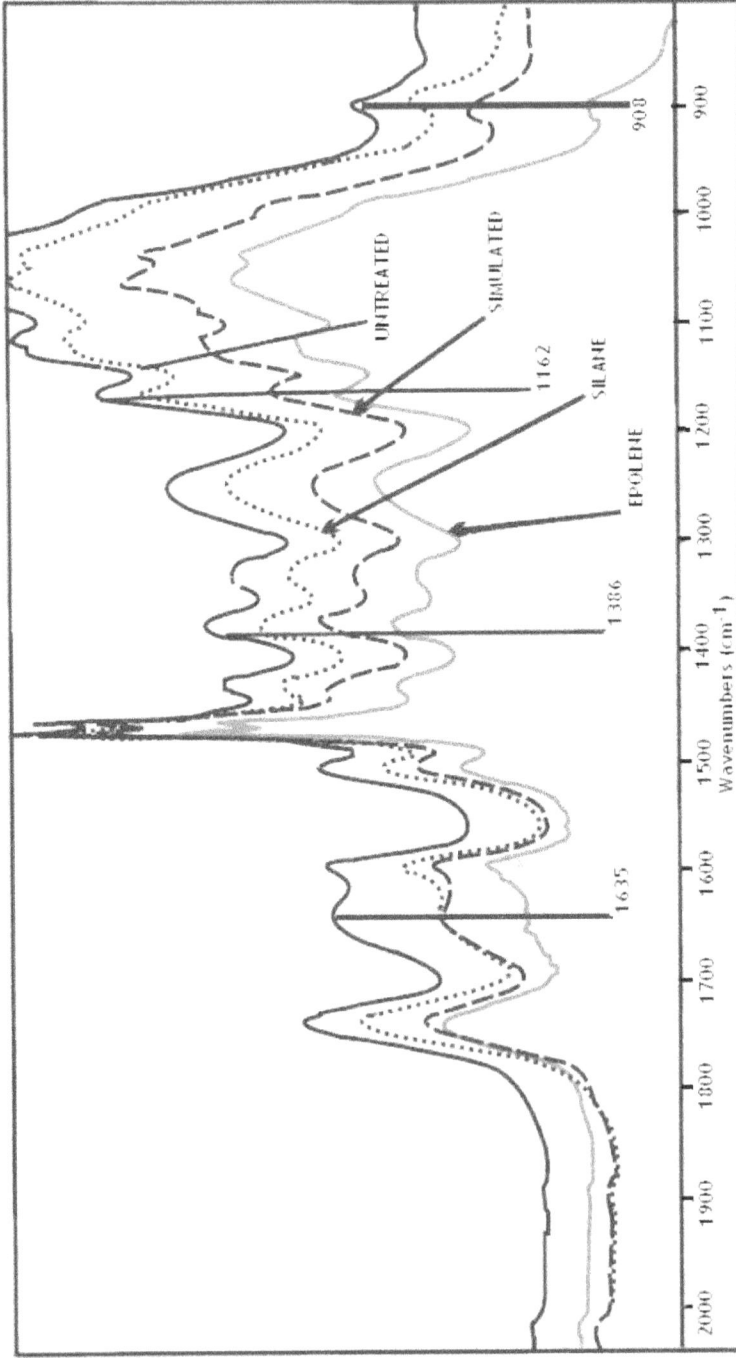

Figure 8.9. FTIR of various types of composites with lignocellulosic reinforcement and HDPE matrix. [Reprinted with permission from X. Colom, F. Carrasco, P. Pagés, and J. Cañavate, *Composites Sci. Technol.* 63, 161 (2003). Copyright © 2002 Elsevier Science Ltd. All rights reserved.]

bending pyran ring at 1465cm^{-1}, CH$_2$ bending (cell) at 1331 cm^{-1}, antisymmetric bridge C–OR–C stretching (cell) at 1162 cm^{-1}, anhydroglucose ring at 1110 cm^{-1}, stretching C–OR (cell) at 1055 cm^{-1}, and antisymmetric out-of-phase ring stetching at 898 cm^{-1} [29,30].

On the other hand, in the HDPE spectrum, it is ncessary to highlight the doublets at 1474–1464 cm^{-1} (–CH$_2$) and 720–730 cm^{-1} (rocking –CH$_2$), the characteristic band at 1368 cm^{-1} (wagging –CH$_2$), as well as the bands of a HDPE at 990 and 910 cm^{-1} (–CH=CH$_2$) [31].

In addition, two special bands, belonging to neither the lignocellulosic fiber nor the HDPE, are observed at 1386 and 1635 cm^{-1}. The spectral band at 1386 cm^{-1} is due to the structural changes, mainly conformational in character, generated by the interaction of the two components. The spectral band at 1635 cm^{-1} corresponds to water absorption, due to the hydrophilic character of the lignocellulosic fiber.

8.2.4.   TGA of Neat Polypropylene and Oil-Palm Wood Flour Filler

Thermogravimetric analysis helps in checking processing temperature by indicating the degradation point of the individual material. One such case study is the TGA of neat polypropylene and oil-palm wood flour (OPWF) filler. The filler degrades earlier than the polypropylene matrix, as shown by the thermogravimetric scans in Figure 8.10 [32]. While quite a substantial amount of degradation (carbonization) of the OPWF has occurred at 380°C, there seems to be hardly any significant degradation of the polypropylene matrix (as shown by the derivative thermogravimetric curve, DTG) up to this point. Thus this temperature may serve as the upper limit of analysis in the subsequent computation for the OPWF content in the composites.

8.2.5.   FTIR of Impregnated Wood Polymer Composite

The major drawbacks of using wood species are their high moisture uptake, dimensional instability, and high probability of biodegradation. These effects are especially pronounced in tropical areas, where wood suffers from exposure to sunlight and high hygroscopicity, which cause swelling and deformation. To overcome these problems and to improve the interaction and compatibility of polymer to the cell-wall component of wood, wood samples were impregnated with methyl methacrylate (MMA) and combined with a cross-linker monomer, hexamethylene diisocyanate (HMDIC). The main purpose of this

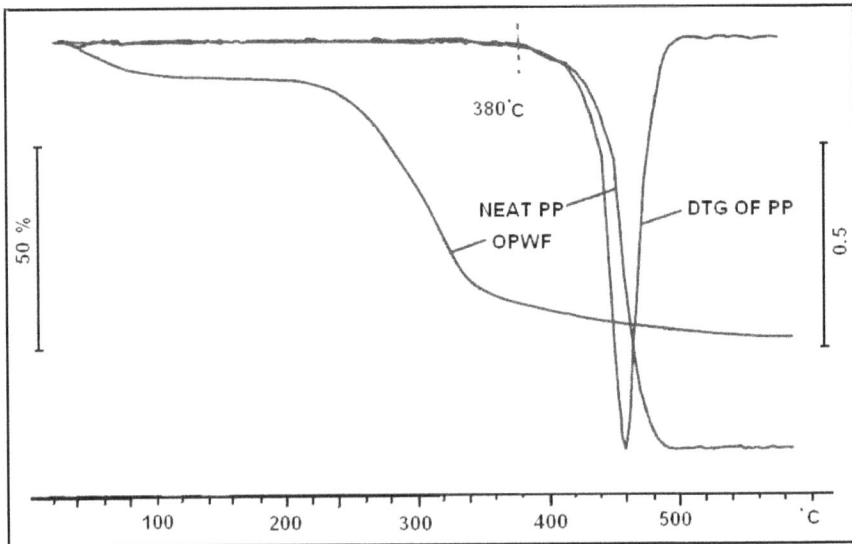

Figure 8.10. Thermogravimetric scans of neat polypropylene and oil-palm wood flour (OPWF) filler. The derivative thermogravimetric curve of polypropylene indicates that most degradation occurs after 380°C. [Reprinted with permission from M. Y. Ahmad Fuad, M. J. Zaini, M. Jamaludin, and R. Ridzuan, *Polym. Testing* 13, 15 (1994). Copyright © 1994 Elsevier Ltd. All rights reserved.]

study was thus to determine the effect of MMA impregnation in the presence of the cross-linking monomer HMDIC on the dimensional stability and mechanical properties of some selected tropical light hardwood composites.

Hexamethylene diisocyanate (HMDIC) is a difunctional reagent which has two reactive functional groups. FTIR spectra of raw wood and methyl methacrylate (MMA), MMA-HMDIC–impregnated wood polymer composite (WPC) samples are presented in Figure 8.11 [33]. The FTIR spectrum of the raw wood clearly shows the characteristic absorption band in the region of 3418, 1736, and 2933 cm^{-1} due to O–H stretching vibration, C=O stretching vibration, and C–H stretching vibration, respectively. From spectrum II, it can be seen that the peak was at 1736 cm^{-1}, which, due to carbonyl stretching vibration, partially disappeared upon impregnation with MMA. The position of the peak at 2933 cm^{-1} (O–H stretching) remained unchanged after incorporation of MMA. On the other hand, the particularly strong O–H stretching absorption band was replaced by a much weaker absorption at 3359 cm^{-1}, and a new carbonyl absorption band developed in the region of 1688 cm^{-1} (spectrum III) as a result of the interaction with MMA-HMDIC. These changes

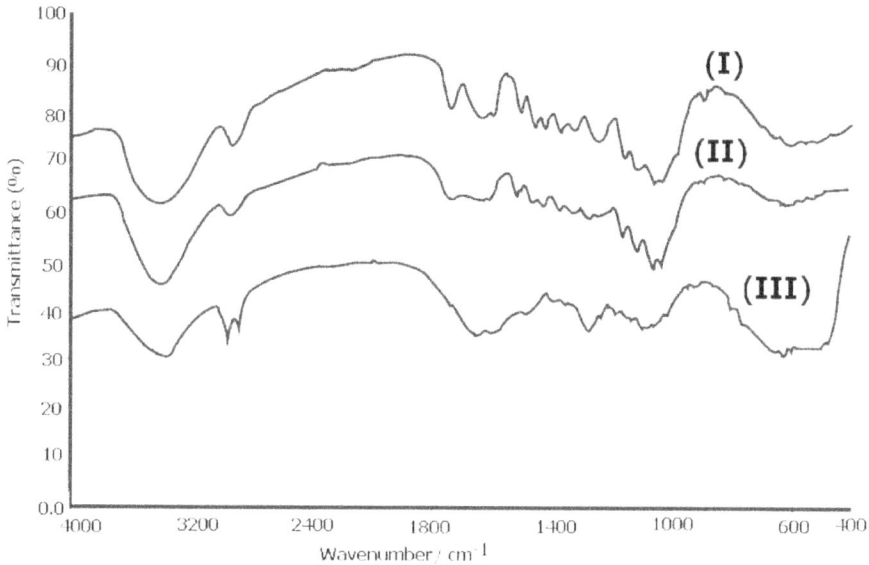

Figure 8.11. FTIR spectra of raw wood (I), MMA-treated wood polymer composite (II), and MMA-HMDIC–treated WPC (III). [Reprinted with permission from Md. Saiful Islam, S. Hamdan, Ismail. Jusoh, Md. Rezaur Rahman, and Z. Abidin Talib, *Ind. Eng. Chem. Res.* 50, 3900 (2011). Copyright © 2011 American Chemical Society.]

were due to the fact that with isocyanate groups of HMDIC, virtually all the hydroxyl groups had been replaced, and the new 3359 cm^{-1} absorption was due to the carbamate N–H bonds as shown in spectrum II. Therefore, it can be confirmed that HMDIC reacted with wood fiber and produced a wood–O–C(=O)–NH–R compound. It can also be seen from spectrum II that the carbonyl band at 1736 cm^{-1} had completely disappeared, and the absorption band of the C–H group had shifted toward higher wavenumbers (2918–2933 cm^{-1}) with narrow band intensity, which gave further evidence of the interaction and cross-linking between wood, MMA, and HMDIC.

8.2.6. Thermal Degradation of Untreated and Alkali-Treated Sesame Husks Composites

The thermal degradation behavior of untreated (USH) and alkali-treated sesame husk (ATSH) composites is shown in Figures 8.12 and 8.13 [34]. The degradation behavior is almost similar up to 400°C. After that, the

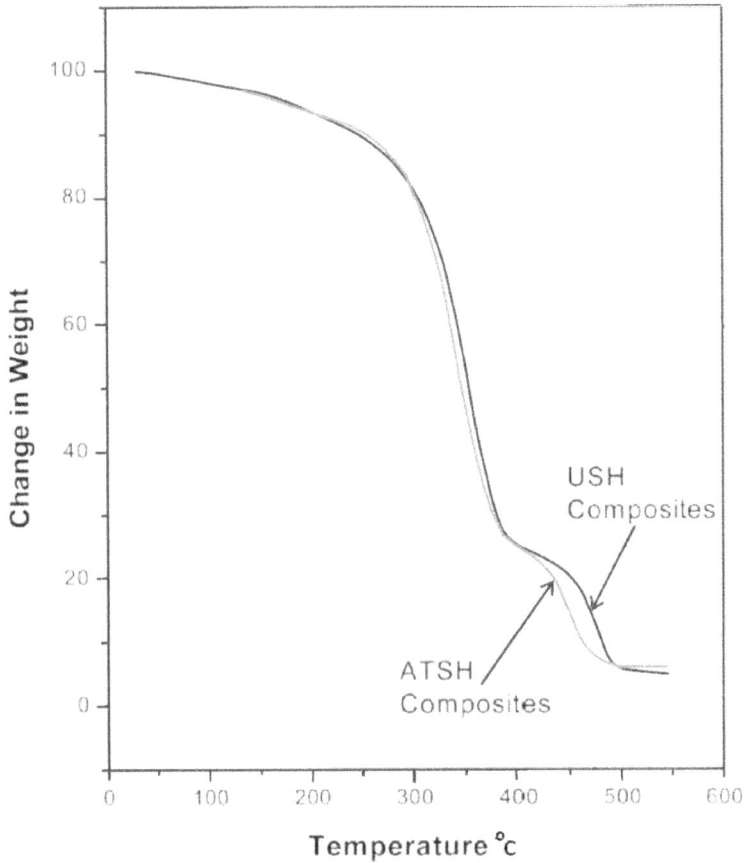

Figure 8.12. Thermal degradation behavior of sesame husk reinforced composites as observed by TGA. [Reprinted with permission from D. Ray, K. Das, S. N. Ghosh, N. R. Bandyopadhyay, S. Sahoo, A. K. Mohanty, and M. Misra, *Ind. Eng. Chem. Res.* 49, 6073 (2010). Copyright © 2010 American Chemical Society.]

degradation peak and the rate of degradation show some differences, which can be attributed to the removal of some lignin from the sesame husk as a result of alkali treatment.

The properties of a wood composite can be significantly improved by the addition of an isocyanate compound to the vinyl or acrylic monomer treating mixture. As noted above, hexamethylene diisocyanate (HMDIC) is a difunctional reagent which has two reactive functional groups. HMDIC modification of wood relies on modifying the predominant wood cell wall polymer by reacting wood hydroxyl groups with a diisocyanate group to

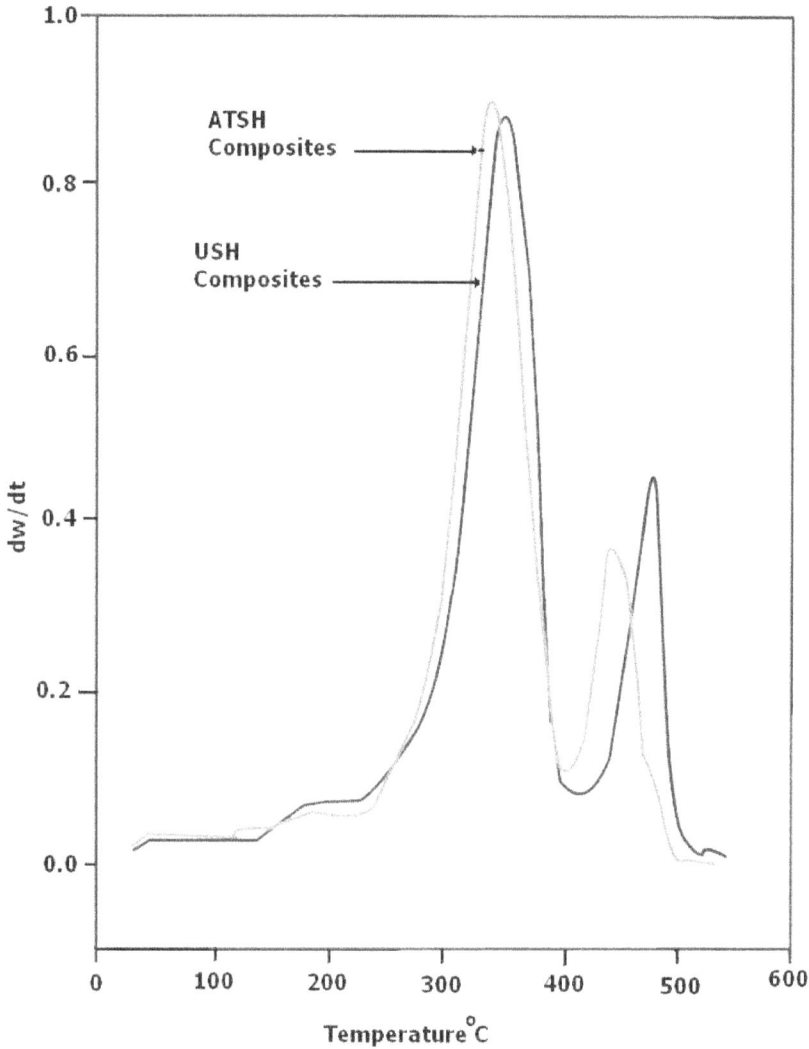

Figure 8.13. Rate of thermal degradation of sesame husk–reinforced composites as observed by TGA. [Reprinted with permission from D. Ray, K. Das, S. N. Ghosh, N. R. Bandyopadhyay, S. Sahoo, A. K. Mohanty, and M. Misra, *Ind. Eng. Chem. Res.* 49, 6073 (2010). Copyright © 2010 American Chemical Society.]

form a wood–urethane derivative. Some researchers consider that the isocyanates also react with accessible –OH groups.

Moreover, the isocyanate group of HMDIC can be exploited for reaction with –OH groups in wood components and for copolymerization with vinyl or acrylic monomers. This reaction can also create new structures in the wood polymer composite that can influence the morphology, crystallization, and also the mechanical, thermal, biological, and other properties of wood. The wood composite products industry has already seen an increase in the use of metal studs for residential construction and plastic/wood as a substitute for wooden decking.

8.3.   Future Trends

Interest in the modification of polymers using blends and composites has been increasing. This trend has inspired polymer scientists to search for and develop high-quality materials with desired properties. The most important factors in creating such are their physical, mechanical, and thermal properties.

8.4.   Conclusion

New instrumental methods have opened up a new era in advances in polymer analytical techniques. Instrumental methods are relatively independent of the field, and history has shown that an interplay of new ideas with multifunctional approaches and advances in technology results in the development of new techniques. Polymer testing is witnessing a glorious period in the breakthrough of inventions and evolution.

With the explosive developments on all fronts, polymer analysis has used selected new technologies with particular emphasis on industrial applications. The impact of modern instrumental methods of polymer testing on science and technology has also been significant. Instrument development has led to precise, automated, and quantitative analysis.

The output of these instrumental methods can provide a wealth of important complementary information. Output of experimental data can not only enhance the level of understanding of problems, it can also save a considerable amount of expensive time. Further, the kind of information that can be obtained has at least as many directions as the number of analytical methods involved.

New instrumental methods for polymers now require multidimensional analytical approaches rather than average properties of the sample. Today, to meet the challenges in analytical methods, hyphenated methodologies are emerging. Polymer separation techniques are being coupled to information-rich detectors or are being interfaced to a second chromatographic system. This is referred to as cross-fractionation or two-dimensional separation.

There are, however, causes of uncertainty that need to be considered in utilizing instrumental methods:

1. Choice of instruments, which differ according to manufacturer and type
2. Quality of the sample and consistency of the method
3. Instrumental setup and sample preparation
4. Data processing software to be used
5. Selection, quality, and concentration of the reagents
6. Effect of contamination during implementation

The main advantage of polymer testing lies in the fact that it employs simple processes that are attractive economically and are less time-consuming than former methods. Polymer testing can provide a clear picture of molecular structures or morphologies involved and may furnish answers to understanding problems related to processing or end products. New instrumental methods will ultimately lead to advances in analytical technologies and sciences.

References

1. W. J. Mac Knight, J. Stoelting, and F. E. Karask, in R. F. Could (ed.), *Multicomponent Component Systems,* Advances in Chemistry Series, 99, American Chemical Society, Washington, DC, p. 29 (1971).
2. D. Deanin, A. Deanin, and T. Sjoblom, in H. Sperling (ed.), *Recent Advances in Polymer Blends, Grafts and Blocks,* Plenum Press, New York, pp. 63–91 (1974).
3. J. A. Manson and L. H. Sperling, *Polymer Blends and Composites,* Plenum Press, New York, p.52 (1976).
4. R. J. Sperber and S. L. Rosen, *Polym. Plast. Technol. Eng.* 3, 215 (1974).
5. D. R. Paul, C. E. Vinson, and C. E. Locke, *Polym. Eng. Sci.* 12, 157 (1972).
6. J. Milgrom, *Incentives for Recycling and Reuse of Plastics,* E.P.A. SW-41 C (1972).
7. D. R. Paul, C. E. Locke, and C. E. Vinson, *Polym. Eng. Sci.* 13, 202 (1973).
8. R. J. Sperber and S. L. Rosen, *Polym. Eng. Sci.* 16, 246 (1976).
9. J. Wypych, *Polyvinyl Chloride Degradation,* Elsevier, Amsterdam (1985).
10. M. N. Radhakrishnan Nair, G. V. Thomas, and M. R. Gopinathan Nair, *Polym. Degrad. Stability* 92, 189 (2007).
11. I. E. McNeill, L. Ackerman, and S. N. Guptha, *J. Polym. Sci. Polym. Chem. Ed.* 16, 2169 (1978).
12. I. E. McNeill and S. N. Guptha, *Polym. Degrad. Stability* 2, 95 (1985).
13. Y. Ward and Y. Mi, *Polymer* 40, 2465 (1999).

14. A. C.-Y.Wang and F. Lam, *Polym. Testing* 21, 691 (2002).
15. M. Bains, S. T. Balke, D. Reck, and J. Horn, *Polym. Eng. Sci.* 34, 1260 (1994).
16. P. Schmidt and B. Schneider, *Collect. Czechslov, Chem. Commun.* 28, 2685 (1963).
17. A. J. Miyake, *Polym. Sci.* 44, 223 (1960).
18. H. Tadokoro, M. Kobayashi, H. Yoshidome, K. Tai, and D. J. Makino, *Chem. Phys.* 49, 3359 (1968).
19. G. M. Venkatesh, R. D. Gilbert, and R. E. Fornes, *Polymer* 26, 45 (1985).
20. J. K. Haken and R. L. Werner, *Spectrochim. Acta* 27A, 343 (1971).
21. G. Socrates, *Infrared Characteristic Group Frequencies,* John Wiley, New York (1980).
22. I. O. Salyer and A. S. Kenyon, *J. Polym. Sci.* A-I 9, 3083 (1971).
23. C.-S. Ha, M.-G. Ko, and W.-J. Cho, *Polymer* 38, 1243 (1997).
24. K. Oksman and H. Lindberg, *Holzforschung* 49, 249 (1995).
25. B. English, P. Chow, and B. S. Bajwa, in R. M. Rowell, R. A. Young, and J. K. Rowell (eds.), *Paper and Composites from Agro-Based Resources,* CRC Press, New York, pp. 269–299 (1997).
26. S. M. B. Nachtigall, G. S. Cerveira, and S. M. L. Rosa, *Polym. Testing* 26, 619 (2007).
27. R. Li, *Polym. Degrad. Stability* 70, 135 (2000).
28. X. Colom, F. Carrasco, P. Pagés, and J. Caňavate, *Composites Sci. Technol.* 63, 161 (2003).
29. R. T. ŎConnor, *Instrumental Analysis of Cotton Cellulose and Modified Cotton Cellulose,* Marcel Dekker, New York (1971).
30. J. Brandup and E. H. Immergut, *Polymer Handbook,* 3rd ed., Wiley-Interscience, New York (1989).
31. D. I. Bower and W. F. Maddams, *The Vibrational Spectroscopy of Polymers,* Cambridge University Press, Cambridge, U.K. (1989).
32. M. Y. Ahmad Fuad, M. J. Zaini, M. Jamaludin, and R. Ridzuan, *Polym. Testing* 13, 15-24 (1994).
33. Md. Saiful Islam, S. Hamdan, I. Jusoh, Md. Rezaur Rahman, and Z. Abidin Talib, *Ind. Eng. Chem. Res.* 50, 3900 (2011).
34. D. Ray, K. Das, S. N. Ghosh, N. R. Bandyopadhyay, S. Sahoo, A. K. Mohanty, and M. Misra, *Ind. Eng. Chem. Res.* 49, 6073 (2010).

Index